T0291805

CAMBRIDGE LIBRARY COLLECTION

Books of enduring scholarly value

Life Sciences

Until the nineteenth century, the various subjects now known as the life sciences were regarded either as arcane studies which had little impact on ordinary daily life, or as a genteel hobby for the leisured classes. The increasing academic rigour and systematisation brought to the study of botany, zoology and other disciplines, and their adoption in university curricula, are reflected in the books reissued in this series.

The Romance of the Apothecaries' Garden at Chelsea

Published in 1928, Drewitt's charming and engaging study traces the origin and antiquity of the peaceful botanical garden in Chelsea. The garden was established in 1673 by the Society of Apothecaries in order to train apprentices to identify the plants used in medicine that, later on, they would be prescribing for their patients. Revised and enlarged for its third edition, the book recognises the special character of this garden, which still teaches students the names and properties of plants, as it did in the time of the Stuarts. Describing the gentle lives of the past naturalists and botanists of the Physic Garden, the study also includes material on the visits of Linnaeus and his pupils. Written with a light touch and full of fascinating anecdotes, the book will appeal to those interested in botany, the history of medicine and the history of early modern London.

Cambridge University Press has long been a pioneer in the reissuing of out-of-print titles from its own backlist, producing digital reprints of books that are still sought after by scholars and students but could not be reprinted economically using traditional technology. The Cambridge Library Collection extends this activity to a wider range of books which are still of importance to researchers and professionals, either for the source material they contain, or as landmarks in the history of their academic discipline.

Drawing from the world-renowned collections in the Cambridge University Library, and guided by the advice of experts in each subject area, Cambridge University Press is using state-of-the-art scanning machines in its own Printing House to capture the content of each book selected for inclusion. The files are processed to give a consistently clear, crisp image, and the books finished to the high quality standard for which the Press is recognised around the world. The latest print-on-demand technology ensures that the books will remain available indefinitely, and that orders for single or multiple copies can quickly be supplied.

The Cambridge Library Collection will bring back to life books of enduring scholarly value (including out-of-copyright works originally issued by other publishers) across a wide range of disciplines in the humanities and social sciences and in science and technology.

The
Romance of the
Apothecaries' Garden
at Chelsea

FREDERIC DAWTREY DREWITT

CAMBRIDGE
UNIVERSITY PRESS

CAMBRIDGE UNIVERSITY PRESS

Cambridge, New York, Melbourne, Madrid, Cape Town, Singapore,
São Paolo, Delhi, Dubai, Tokyo, Mexico City

Published in the United States of America by Cambridge University Press, New York

www.cambridge.org
Information on this title: www.cambridge.org/9781108015875

© in this compilation Cambridge University Press 2010

This edition first published 1928
This digitally printed version 2010

ISBN 978-1-108-01587-5 Paperback

THE
ROMANCE OF THE
APOTHECARIES' GARDEN
AT CHELSEA

Cambridge University Press
Fetter Lane, London
New York
Bombay, Calcutta, Madras
Toronto
Macmillan
Tokyo
Maruzen-Kabushiki-Kaisha

CHELSEA PHYSIC GARDEN FROM THE RIVER
Showing Cedars of Lebanon planted in 1683. Both lived until 1878—the survivor until 1903
From a lithograph of a water-colour drawing by James Fuge, circ. 1830

THE
ROMANCE OF THE
APOTHECARIES' GARDEN
AT CHELSEA

by

F. DAWTREY DREWITT, M.A., M.D.

FELLOW OF THE ROYAL COLLEGE
OF PHYSICIANS

*A Garden . . . is the purest of human pleasures;
it is the greatest refreshment to the spirits of man.*

Francis Bacon

THIRD EDITION

CAMBRIDGE
AT THE UNIVERSITY PRESS
1928

First Edition   1922
Second Edition 1924
(Chapman & Dodd, Ltd.)
Third Edition   1928
Revised and enlarged
(Cambridge University Press)

PREFACE

A short time ago the writer was asked to represent the Royal College of Physicians on the Managing Committee of the Chelsea Physic Garden—now controlled by Trustees of the London Parochial Charities.

The request was readily complied with. It was an opportunity of learning the long and interesting history of the Garden, and of reading the records of its public-spirited supporters, and of its rare trees and flowers.

The story may interest some who were unaware of the existence of the Garden.

Many must have found, with the writer, an absorbing pleasure in exploring some minute fraction of the great human past—of understanding something of "yesterday, its aim and reason." It has all the novelty of a journey in an unknown country.

This short journey has been exceptionally pleasant; for it has taken the writer among men of gentle and attractive lives—the old naturalists and botanists of the Physic Garden—men who lived near Nature—devoted to their dear mother Earth.

F. D. D.

July 1922

PREFACE

TO SECOND EDITION

The second edition—revised and enlarged—
contains a letter kindly sent to the Author by
the Revd. W. E. Layton. It was written by
Gronovius, a friend of Linnæus', to Mr Layton's
ancestor, Philip Miller, the well-known curator
of the Garden.

The letter shows that it was, without doubt,
the Apothecaries' Garden which brought Linnæus
to England.

Many additions have also been made to the
chapter describing the trees. But it must not
be thought that there are trees now in the
Garden which cannot be found elsewhere; for
Kew Gardens have supplied London Parks with
trees of many kinds—Battersea Park, in 1856,
receiving no less than five thousand, with the
present happy result.

This new edition also contains further notes
on Linnæus, and on his two pupils, Kalm and
Fabricius, who came to Chelsea.

It is a pleasure, always, to a writer to find
that others are interested in subjects that have
interested him.

F. D. D.

January 1924

PREFACE

TO THIRD EDITION

In this edition, further revised and enlarged, many additions have been made. There are two new chapters.

Evidence is examined for the disputed tradition that Linnæus, having walked from Chelsea to Putney Heath, fell on his knees before the yellow gorse.

There is an account of the end of his life—of the sale of his Collections—of their arrival in Chelsea.

Letters—hitherto unpublished—written by Sir James Smith, the purchaser of the Linnean Collections, to William Jones of Manor Street, suggesting a "Linnean Society" are given.

There is an account of the formation of the Linnean Society, and of some of its early members.

The frontispiece, carefully reproduced from an old lithograph, showing the well-known cedars, is the work of Mr Emery Walker.

F. D. D.

May 1928

CONTENTS

ILLUSTRATIONS

INTRODUCTION

The term "Physic Garden"—the official name for the Apothecaries' Garden—is a little misleading.

Although for a few years a small part of the Garden supplied herbs to be used in the laboratory at Blackfriars—and although the chief reason for the study of plants at that time was to discover their medicinal qualities—it will be seen that the Garden was founded, not for the production of drugs, but for the advancement of botany.

The word "physic" had then a wider meaning than it has now. It could be used in its original sense of pertaining to physical (*i.e.* natural) science. A "Physic garden" was a scientific garden, and, like physical strength and physical courage, not necessarily connected with drugs.

In the 17th century the two old botanic gardens of England—those of Oxford and Chelsea—were both styled "Physic" Gardens. Evelyn, at Oxford for Commemoration, tells us that he went to "the Physic Garden, where the sensitive plant was shewed us for a great wonder. ... There grew canes, olive trees, rhubarb, ... besides very good fruit, which when the ladies had tasted, we

returned to our lodgings." It was not a drug-
producing garden.

The same may be said of the Garden of the
Princess Dowager of Wales at Kew, which was
made into a "Physic" Garden in 1760, and was
the small beginning from which the Royal Botanic
Gardens grew.[1] The "Physic Garden" at Kew
House was simply a scientific collection of her-
baceous plants arranged by a pupil of Philip
Miller's according to the new Linnæan system.

There the name "physic" was soon lost, and
the Oxford Physic Garden had already become
the "Botanical Garden." But the old label stuck
fast to the Chelsea Garden—naturally—for it
was the garden of the Apothecaries.

The books on which the following history
chiefly depends are Faulkner's *History of Chel-
sea*, 1829; Field and Semple's *Memoirs of the
Botanic Garden at Chelsea*, 1878, and Barrett's
History of the Society of Apothecaries, 1905.
But other works of interest have been consulted.
The British Museum Library has been thankfully
used, both at Bloomsbury and at its Natural
History branch in Kensington.

A "Life" of Sir Hans Sloane (to whom the
long existence of the Physic Garden is chiefly

[1] A. W. Hill, F.R.S., *Annals of Missouri Botanical Garden*, 1915.

due) occupies forty-five pages of Faulkner's *History of Chelsea*, and thirty of Pulteney's *Progress of Botany*. There is a concise account of him in the *Dictionary of National Biography*, by Sir Norman Moore, who lately occupied Sloane's Chair as President of the College of Physicians. There is a long but readable biography in *By Chelsea Reach*, by Reginald Blunt; another in Munk's *Roll of the College of Physicians*; and the diary of Sloane's critical caretaker, Howard, is quoted in *The Greatest House in Chelsea*, by Randall Davies.

Copies of all editions of *The Gardener's Dictionary*, written by Philip Miller at the Physic Garden, are in the Natural History Museum in Cromwell Road; also Sir Joseph Banks' great collection of dried plants.

Minute books of the College of Physicians give the information that John Gerarde agreed to be superintendent of the first botanic garden of the College—in 1587.

THE APOTHECARIES' GARDEN

Chapter i

IT is indeed a romance that, in spite of the
continual destruction of its old life by modern
London, spreading like a flood, submerging and
changing, beyond recognition, the face of all the
country it touches—in spite of perennial calls
for abandonment—the peaceful Garden of the
Apothecaries—a living book on Botany—should
still be teaching its students the names and nature
of plants, as it did in the days of the Stuarts.

It is fitting that it should be so, for botany is
a very ancient science. It was man's first lesson
in Nature study. His very existence depended
on his learning it.

Poor primitive man, as he crawled out of his
cave at sunrise, with no anxiety for the morrow,

and but slightly sensitive to cold and pain, had
to learn by hard experience that, although the
blackberries which grew on the prickly bushes
were good eating, the smooth, black berries of
the deadly nightshade would kill him—although
the rabbits nibbled its leaves without hurt. The
lower animals had taught him much—they had
learnt their lesson before him—but their teaching
was by no means infallible.

He had to discover that the roots and leaves
of wild cabbage were wholesome, but that the
roots of the monkshood, close by, would stop
his breath—that the tempting fruit, like an
orange, lying on the thirsty desert sand, which
his descendants would some day call "bitter
apple," was poisonous, although he had seen the
wild rock pigeons peck it, and had found the
seed in their crops.[1] He had to learn that he
might eat the luscious red berries on the Yew
tree, for which the missel-thrushes scrambled in
the autumn—but not its green seeds or fresh
twigs—that grass seed[2] was wholesome, and

---

[1] The writer found seeds of Colocynth in the crop of a Rock
Dove shot in South Algeria.

[2] Wheat, barley and oats have been so long cultivated, that it
is difficult to trace the wild grasses from which they came. De
Candolle (*Origin of Cultivated Plants*) believes that wild wheat
exists in Mesopotamia. The wild ancestors of oats and common
barley are said to be found in Western Asia.

that, just as the wild oxen ground it with their great teeth, he could grind it between stones into meal.

And so the cave man became, of necessity, a field botanist—a better one than many a modern Londoner.

Time went on. The medicine-men made drawings of herbs on clean calf-skins, to show their fellow men which plants were wholesome, and which were poisonous; and they wrote down on the skins the virtues of healing they believed each herb to possess.

So the medicine-men became the teachers of botany, and the early descriptions of plants were written by them—from the great book of Dioscorides, Army Surgeon to the Roman troops about the time of Nero—of which a copy, written in Greek capitals, is to be seen in Vienna, with Arabic notes to its illustrations, probably made in Constantinople in the ages when Europe slept—down to the herbal of our own John Gerard, who practised medicine in Holborn in Shakespeare's day—a book full of well-drawn, simple woodcuts, showing the infinite variety and beauty of the outlines of leaves and flowers.

So the Apothecaries were rightly lineal descendants of those old medical botanists, and their botanic garden at Chelsea is linked with the whole history of botany during the last two and a half centuries.

The Apothecaries' Society has an interesting history. It had a stormy beginning. It was born when James the First was King—his own child—and the King stood by it like a father in all its early struggles. In 1617 it broke away from the great Grocers' Company, and the reasons for its declaration of independence are given in its charter, which runs thus:

JAMES, by the grace of God, King, Defender of the Faith. . . . To all whom these presents shall come greeting. Whereas . . . very many Empiricks and unskilful and ignorant Men . . . do abide in our City of London . . . which are not well in-structed in the Art or Mystery of Apothecaries, but . . . do make and compound many unwholesome, hurtful, deceitful, corrupt, and dangerous medi-cines and the same do sell . . . and daily trans-mit . . . to the great peril and daily hazard of the lives of our subjects . . . We therefore . . . weighing with ourselves how to prevent the endeavours of such wicked persons . . . thought necessary to disunite and dissociate the Apothecaries of our City of London from the Freemen of the Mystery of Grocers . . . into one body Corporate and

Politic... to whom in all future times the management of those inconveniences might be given in charge and committed... after the manner of other Companies.[1]

So the order came to the Apothecaries to leave the Company of the Grocers, and form themselves into a separate City Company.

But the birth of the Apothecaries' Company was no simple matter. It was full of trouble. Like Pharaoh of old, the Grocers refused to let the people go. The Apothecaries had no thought of spoiling the Grocers. They were taking neither endowments nor silver; but the Grocers disliked the prospect of having their numbers reduced, and did all that was possible to hinder the exodus. Some Apothecaries, too, seemed to prefer the flesh-pots of the Grocers, and murmured against their leaders, who were taking them into the wilderness, where they had no habitation, and where some day they might be heavily taxed to find one.

But King James stood by them. And so, with little money and no home, the Apothecaries formed themselves into a City Company, and

[1] The whole charter is quoted in Barrett's *History of the Society of Apothecaries*, and occupies no less than 21 quarto pages.

undertook the difficult duty of improving both
the quality of the drugs, sold to His Majesty's
subjects, and the qualifications of the vendors.

The Grocers in the meantime were not idle.
In 1624 they induced the Mayor and Corporation
to petition the King to revoke his charter, and
the House of Commons to support the Lord
Mayor's appeal. The alarmed Apothecaries hastily
summoned a meeting, and a call of £1 a head
was made to defray expenses of counsel; but it
was agreed to accept a fate which appeared inevitable, and return (under certain conditions)
to the Grocers. An end to the Apothecaries'
Company seemed in sight.

But the Stuarts, whatever their failings, were
on the side of science. King James himself met
the Mayor and Corporation—told them that he
gave the Apothecaries their charter "from his
own judgement for the health of the people,
knowing that grocers are not competent judges
of the practice of medicine," and that his inten-
tions would be made known to the Speaker of
the House of Commons—informed the House
of Lords that the establishment of the new
Company was "a general good," and sent a
warrant to the Apothecaries to proceed in the
due execution of their charter "notwithstanding

the proceedings in the House of Commons." The
Company was saved.

Autocracy would be an excellent form of
government if autocrats were always wise. It is
unfortunate that the possession of uncontrolled
power should corrupt its possessor; and that the
unopposed despot should slide so easily down
the smooth road to Avernus.

King James had thus made the position of the
tottering Apothecaries' Company firmer than
ever, but the civic authorities could hardly be
expected to welcome the new-comers. It was not
until seven years afterwards that the Lord
Mayor sent them an invitation to be present
at St Paul's, and hear the sermon on Christ-
mas Day, together with the other City Companies.

Peace was made, and civilities were exchanged.
Barrett quotes a long entry in the Company's
books relating to the happy event: "After the
sermon the Lord Mayor and the Aldermen came
in a most noble and courteous manner and saluted
the Company." The Apothecaries presented the
Lord Mayor with "a tun of wine." Next year
they took part in the procession on Lord Mayor's
Day in a barge hired for the occasion, but under
banners of their own.

Then their wanderings in the wilderness came
to an end. Cobham House,[1] on the bank of the
Thames, where the little Fleet river joined it at
Blackfriars, was purchased from Lady Howard
of Effingham. It had waste land and tenements
all the way down to its landing stage on the
Fleet. A road called Water Lane led to the
house from Ludgate Hill, and on to the Thames
at Blackfriars Stairs (under the present railway
bridge).

There was a road from Cobham House across
Water Lane, down the bank of the Fleet to the
landing place—a steeper descent than it is at
present; for New Bridge Road, to which it now
leads, was afterwards built above the Fleet,
which still flows underneath it.[2] A motor omnibus
to-day passing under Holborn Viaduct on its way
to Blackfriars Bridge, by Ludgate Circus, would
run along and above the imprisoned Fleet river—
up which barges once found their way to Holborn.

Almost next door to Cobham House was a
small theatre used by a theatrical company in
the winter, when their summer theatre, "The
Globe," across the water, was found too cold and

[1] In 1600 Lord Cobham had entertained Queen Elizabeth at a
masque at Cobham House.—*London Past and Present*, Wheatley.
[2] The arching of the Fleet was completed in 1765.

too much exposed to the weather. Blackfriars
theatre was roofed all over, well lit with tallow
candles, and, on a winter's evening, a pleasanter
place than "The Globe" for both actors and
audience.

Many of the Apothecaries must have known
it twenty years before, in the lifetime of one of
its managers, who wrote many plays, put life into
all his stories, and whose company had often performed before King James, and even before the
old Queen, who "to his lays opened her royal ear."[1]

Although Blackfriars theatre was not quite
what it had been, the Apothecaries' apprentices
must still have enjoyed their laugh at Falstaff,
and have come away inspired by Henry the
Fifth. But the theatre was slowly coming to an
end. The frivolous side of play-acting had perhaps
grown, and, on the other hand, men's minds had
become more anxious and serious. A few years
later—in 1642—an order came to close the play-
house.[2]

Cobham House, with its land and smaller
houses, cost the Apothecaries £1000—not an

---

[1] Chettle. *Life of William Shakespeare*, by Sir Sidney Lee,
p. 376.

[2] The Lord Mayor had already tried, ineffectually, to close the
theatre, owing to the crowd it occasioned in Water Lane.—
London Past and Present, Wheatley.

extravagant sum. Arrangements with tenants
were soon made, repairs and alterations finished,
and in December, 1632, the Apothecaries' Company—Corporation, Freemen and all—met there
and rejoiced on entering into their promised
land.

Higher qualifications were now required for
membership. Physicians were asked to attend
the examinations. Stewards were appointed to
arrange botanical excursions. Bad and dangerous
drugs exposed for sale were seized, and burnt at
the Hall.

One of the members, Thomas Johnson, of
Snow Hill, further up the Fleet, had finished
his enlarged edition of Gerard's *Herball*, and
was thanked for a copy he presented. *Johnson's
"Gerard"*—a great book, well-known to botanists
—much used in country houses, in the 18th
century, as a book on botany and weird domestic
medicine—in the 19th century, as a book of
designs for art needlework.

Thomas Johnson had exhibited in his shop
window in Snow Hill the first bunch of bananas
seen in London.[1] He had received them from his

[1] This account of a bunch of bananas coming to the President
of the College of Physicians from the Bermudas is interesting.
Bananas had been cultivated in the Malay Islands, India and

"much-honored friend, Dr. Argent, President of the Colledge of Physitions." They had come all the way from the "still-vex'd Bermoothes" in a sailing vessel. Johnson hung them up in his window on April 10, 1633, and they lasted until June, when they were "soft and tender."[1] The air of London could not have been so very in-sanitary in those days—even on the banks of the Fleet!

Johnson had his bananas carefully drawn and engraved. He cut small slices of them, and found that they had a pleasant taste, and no seeds. Little he thought of the millions of Londoners who would some day follow his example!

China from time immemorial; but the question whether they existed in America in prehistoric times has been much disputed.

De Candolle, in *Origin of Cultivated Plants*, 1884, p. 304, after summing up all the evidence, decides that until the Spanish Conquest the New World had no bananas.

Little was known of the Bermudas until Sir George Summers was wrecked there in a tempest in Shakespeare's time (1609), and by extraordinarily good fortune escaped with all his com-panions to Prospero's Island. So this bunch of bananas at first sight suggests that they were indigenous there.

But a *Historye of the Bermudaes*, of about 1631, among Sir Hans Sloane's MSS, records that a "Great abundance of fig trees, plantans (*i.e.* bananas), vines and orange trees," had been *intro-duced*.

[1] "The pulp, or meat was very soft and tender, and it did eate somewhat like a muske-Melon. I have given you the figure of the whole branch, with the fruit thereon, which I drew as soon as I received it." *Johnson's "Gerard,"* p. 1515.

Johnson was a good fellow—companionable, and fond of making botanical excursions into the country with those of like mind. The first local list of wild flowers published in England was made by him.

Of four of these expeditions accounts were published—one a journey into the fields of Kent —*"Iter in agrum Cantianum"*—by ten companions—in 1629; another to Hampstead Heath —*"Ad ericetum Hamstedianum"*—with a list of the flowers met with. Wild Bugloss was then growing on the dry ditch banks about "Piccadilla,"[1] and Belladonna in Islington. Later on two excursions were described under the title of *"Mercurius Botanicus"*—the Botanical Tourist —one a trip to Oxford, Bath, Southampton, and the Isle of Wight—and another to Wales and Snowdon, in search of wild flowers.

These "herbarizing" parties became organized institutions of the Apothecaries' Company. The present County Field Clubs are no doubt their direct descendants—possibly, too, the great natural history excursions led by Linnæus in Sweden.

[1] Not the present street. A draper who had made a fortune out of Piccadillas (stiff lace collars) built a house near St Martin's in the Fields. It was called " Piccadil Hall " and the neighbourhood " Piccadilla." The road to Knightsbridge was named " Piccadilly Street " fifty years later.

TITLE-PAGE OF JOHNSON'S EDITION OF *GERARD'S
HERBALL* IN THE LIBRARY AT APOTHECARIES'
HALL

*Johnson was an original member of the Apothecaries' Company.
On the left of Gerard's portrait is seen a bunch of the first bananas
brought to London*

Of Gerard's *Herball* Johnson made an entirely new book; 800 descriptions of new plants were added to it,[1] with 700 woodcuts. Many of Gerard's mistakes were corrected. But the book was still to bear Gerard's name; and, on a well-engraved title-page, Johnson—modest gentleman that he was—instead of his own portrait, put Gerard's in Elizabethan ruff—would that there was as good a one of Shakespeare!

The accompanying photograph of the title-page, reduced to one-fourth its size, is taken from a copy of the book in the library of the Apothecaries' Society. It is not the one Johnson gave. That one must have formed fuel for the Great Fire, when molten lead from the roof of St Paul's flowed down Ludgate Hill like a lava stream.

Ceres with wheat and Indian corn, and Pomona with apples and pears, are at the top of the plate. They have their place in a "Physic" herbal, as well as the flowers with their butterflies and dragon-fly at the foot. The bunch of bananas which Johnson hung in his shop window, and tasted, is on Gerard's right, and has not yet been placed among the gifts of Pomona.

[1] Eighty-eight plants new to Middlesex were described. Trimen and Dyer, *Flora of Middlesex*, p. 371.

Johnson was given the freedom of the Apothecaries' Company, and an M.D. degree of Oxford. The outbreak of civil war stopped all botanical studies. Johnson joined the Royalists—took an active part in defending Basing House—was made Lieut.-Colonel—fought bravely—was wounded and died, and left the world the poorer.

A writer of the time said of Johnson: "His worth did justly challenge funeral tears ... he was no less eminent in the garrison for his valour and conduct as a soldier, than famous through the kingdom for his excellency as a physician."[1]

So war ever, among civilized races, where the sickly, the feeble in body or mind, are protected, and those sound in mind and healthy in body are exposed for slaughter, brings about the destruction of the best, and ensures the survival of the unfittest, and deterioration of the race.

To all, especially to those who see in the long history of the world ordered growth and evolution from lower to higher forms of life—in man as in the lowest—this destruction of all that is excellent—fine character, wisdom, learning, strength—and the survival of what is weak, and often criminal, is depressing. In the struggle for

---

[1] *Description of the Siege of Basing Castle*, Oxford, 1644.

existence Abel dies, and Cain and his descendants seem the fittest to survive.

But the great mills grind slowly; and time is infinite; and these disasters must be looked upon as only local checks—even when they involve nations. The tide advances, though we see the retreating wave.

The tree Igdrasil of the old Norse legend—the tree of life, and of wisdom—still grows, and covers the earth; though branches decay, and Nishud, the dragon, gnaws the root; for the chief root goes down to the sacred fountain of Urd, where the gods meet in council.

So Johnson died, and was buried; and no man knows his grave. But the flowers of the Johnsonias[1] are his wreath; and his *Herball* is a monument more lasting than stone.

Other members of the Company now helped with donations. One of them offered as much as £500 towards a laboratory on the "waste land."

A private pew, with keys for the Master and Wardens, was reserved for the Company in Blackfriars church. Dinners in future were to be held in the Apothecaries' great Hall instead of in taverns.

---

[1] A group of lilies.

Little they recked of the trouble to come,
when that great laboratory, the world—ever at
work—ever destroying and creating anew—
would be in full ferment round them; when
England would be torn in pieces by civil war;
when the "Great Sicknesse" would haunt London
with its "death carts," collecting the unburied
bodies—apothecaries' and all—to pile them in
pits like dead dogs; when their new Hall, their
library, and the private houses of their members
would be destroyed in the Great Fire.

Financial troubles meantime pressed heavily
on the Company already in debt. In 1635 came
the demand for money to provide a fleet to help
Spain against Holland—"ship money." Five
years later £300 was wanted from them—part
of the City's compulsory "loan" to Charles I. For
this they were compelled to let the Hall.

Civil war began in 1642, and the City heard
with alarm that the King's troops had reached
Hounslow. Parliament then demanded a loan
to meet emergencies, and the City money for
the repair of London Bridge. The Apothecaries
paid their share, and just managed to save their
small collection of plate. In 1660, at the Restoration, they were required to contribute towards
the City's present to the King.

Then followed the Great Plague, and after the Plague the Great Fire. Their Hall disappeared, and many of their members' houses. There are scanty records of these years. It was not a time for keeping careful minutes. The plate was saved, but no books.

The Fellows of the College of Physicians close by, in Amen Corner (now houses for the canons of St Paul's), were more fortunate. They saved 140 folio volumes, besides valuable manuscripts; and Lord Dorchester, a Fellow of the College, to repair their losses, presented them with his great library, including the Wilton Abbey Psalter of 1250, Caxton's first book printed in English, and the first printed Homer—all to-day, with the exception of one scorched volume, in excellent condition.

They saved pictures, too, and the great silver mace, and 200 years afterwards were able to recover a piece of plate carried off by burglars during the Plague.

But the College of Physicians building was destroyed, with the great museum and library, where Charles II heard a lecture on anatomy—near Stationers' Hall. Fortunately the College[1]

[1] In 1614 the College of Physicians had taken a lease of house and lands at Amen Corner, from the Dean and Chapter of

had a small botanic garden between Amen
Corner and the Old Bailey, which must have
been of use in the rescue of books and plate.

The Barber-Surgeons' Hall in Monkswell
Street also had a garden, which saved their old
building; though the great Holbein painting of
Henry VIII, with Butts, the King's physician
(who, Shakespeare says, pointed out Cranmer in
the crowd), was carried into it for safety.

But though Apothecaries' Hall vanished, the
Company survived. They sold the plate, and
let all the burnt tenements on a building lease;
and, with the smell of fire in their nostrils, put
a clause in the lease that no blacksmith or tallow
chandler should ever be tenant of theirs.

The rebuilding of the Hall began at once—
and its old oak staircase was fashioned by Mr
Young, a freemason, partly for his own pleasure
in the work, partly as rent for land he occupied.[1]

St Paul's; and had given up their house in Knight Rider Street,
and their first Garden.

It is not generally known that John Gerard (Johannes Gerarde)
agreed, on October 6, 1587, to superintend that Garden, and keep
it full of all kinds of plants (... "omni fere herbarum variorum
genere refertum tueri").

[1] This hurried rebuilding of London prevented Sir Christopher
Wren's great scheme for a new London, with convenient streets,
from being carried out.

Chapter ii

THE engaging study of botany—not so scientific at that time as it is to-day, but every bit as enjoyable—depended for information on three sources—herbals (*Johnson's " Gerard,"* and others), collections of plants (pressed, dried, fastened to sheets of paper, and labelled), and, best of all, excursions into the country with recognized teachers in search of growing plants —there was more to be learnt from living things than from dead ones.

But some plants and trees were not to be seen on any excursion. So the Apothecaries, in spite of the debt on the new Hall, and some still unpaid debts on the old one destroyed by the fire, set to work to find a garden where they could cultivate rare plants, and sow seeds now coming in from foreign lands.[1]

[1] The Apothecaries also wanted a convenient spot for a barge. That was one reason for taking a lease of the garden. But rather

A garden was found—a plot of some three and a half acres—in the pleasant riverside village of Chelsea. In 1673 they obtained a lease of it from Charles Cheyne, afterwards Lord Cheyne, for an annual rent of £5.

Chelsea at that time was a country manor. It had its cornfields, pasture, common land, and its village by the water. It was bounded on three sides by rivers—two of them small streams. But even small rivers make efficient boundaries—landmarks no neighbours can remove.

The largest — the Westbourne — formerly flowed by Westbourne Terrace into Hyde Park, spread out into the Serpentine, dipped under Knight's Bridge by Albert Gate, appeared again at William Street, passed down by Lowndes Square and Cadogan Place, parallel with Sloane Street, into the Thames at Chelsea Bridge. Trees grew on both its banks, especially about Sloane Square.

As late as 1809, though growing smaller from the gradual draining of the land, it was able to overflow its banks, flood houses, and convert

more than a year afterwards (Barrett records) "ground for a barge-house was taken from Sir John Sheldon on a lease of 51 years." Possibly near Blackfriars.

lower Chelsea into a great lake; so that those who wished to go from Chelsea to Pimlico had to cross over in boats. Faulkner, the invaluable historian of Chelsea, saw the flood, and describes it as an "awful visitation"—as no doubt it was.

The Westbourne to-day passes harmlessly underground, imprisoned in a huge iron tube, which can be seen over the heads of passengers waiting for trains at Sloane Square station— and so on to the Thames near the grounds of Chelsea Hospital.

But for a few yards the old river still flows in the open air, passing out of the Serpentine through a pleasant dell with grass banks, yellow iris, water lilies and moorhens, before disappearing under Rotten Row.

In 1673 the great Military Hospital with its wide open grounds—destined to be a future neighbour of the Physic Garden—had not yet been planned by Wren. Sir Stephen Fox and Evelyn had not obtained the King's consent.

The long King's Road was being finished. It was made for Charles II, who wanted a direct way from Whitehall to Hampton Court Palace, where French gardeners were busy laying out the gardens.

Starting from the west gate of St James's Park and passing the Mulberry Garden, lately planted by James I (now Buckingham Palace Garden), it curved to the right, and leaving some wet ground about Eaton Square on the left, and ponds in Belgrave Square on the right, it turned to the south about the west end of Eaton Place.

At Sloane Square it crossed the Westbourne, replacing the old footbridge, of evil reputation for robberies, with one strong enough to bear the King's coach.

Here it entered the Manor of Chelsea, and being on firm ground, was carried on in a straight line through the fields—dividing Chelsea in two—and so up to the old ferry which was working between Fulham and Putney when Domesday Book was being written.

It was well made, gravelled, and maintained at the expense of the Crown. The King must have found it an easier road than the old one through the rowdy village of Knightsbridge and the puddles and ruts of the Hammersmith Road.[1]

The plot the Apothecaries had taken was an excellent site for a garden. Cultivated fields in

[1] Kensington was almost isolated by the condition of the road.

East Chelsea, the uninhabited district beyond
(now Belgravia and Pimlico), with its meadows
and ditches, the Tothill fields, and St James's
Park separated the Garden from smoky London.

The Thames alongside kept it open to the
south. Every high tide brought rich river water
for the plants—not unpleasantly rich, as it afterwards became, for the water was clear enough
for angling. Many kinds of coarse fish were to be
found in Chelsea Reach, and the nets had good
hauls of salmon in the spring, in spite of the
poachers who destroyed the young fish with
illegal nets.

There was a creek, too, for the new barge, and
a boat-house in the Garden itself. A recess in the
south-east corner of the Garden marks, to-day,
the spot where the Company's barge, as well as
two other barges, for which the owners paid rent,
could be housed. This recess, and the old river wall
in the Garden, now some way from the river, and
buried, almost to the top, by the raised ground,
show how much the Embankment—made in
1874—gained on the mud-banks formerly left
bare at low tide.

The barges housed in the Garden were not
the London barges we all know—the "lighters"
and the sea-going barges, which float down the

"London river" to the North Sea, with their great sails set before a westerly wind.

The Company's barge was a four-oared rowing boat, with a room, like the cabin of a gondola, in the stern, decorated with flag and banners.

It was of modest size. Sir Thomas More's barge, on which the last sad journey from Chelsea to Lambeth was taken, and many other barges, had eight oars—sometimes more. The river at times would be full of them; for the Thames was a great highway, like the Strand and Cheapside, with the advantage that the highway required no repairs, and was not liable to obstructions. Barges and boats were its cabs and carriages—the watermen its cab-drivers and coachmen.

A Lord Mayor's procession on the river must have been full of life and colour, and not altogether unlike a festival on the grand canal at Venice.

Steamboats on the river, and improved roads for coaches, must have put an end to Thames pageants. Happily the procession of the "eights" at Oxford survives.

The watermen who rowed these barges were a sturdy race, and made good recruits for the navy. Thomas Doggett, Irishman, actor, convinced Whig, founded an annual prize for them,

CHELSEA PHYSIC GARDEN

Kœlreuteria Paniculata (*rare Chinese tree*) *near Swan Walk entrance*

to be rowed for every August, in commemoration
of the Accession of George I—an orange-coloured
coat with the Hanoverian Horse in silver on it,
as a badge. The race formerly finished at the old
Swan Tavern at the corner of the Apothecaries'
Garden; but to-day it is continued some yards
further up the river.

Swan Walk, which bounds the east side of the
Garden, was a footpath leading to the house.

It was at the Swan Tavern that a jaunt of
Pepys' came to an abrupt end. He tells us that
in April, 1666, he drove to it with two ladies
and two children, "thinking to have been merry,"
but found the house shut up for the plague. He
says: "We turned back with great affright, I for
my part in great disorder."

Bishop's Walk bounded it on the west.

George Morley, Bishop of Winchester, was
living close by—a man of fine character—friend
of John Hampden, and of Edward, Earl of
Clarendon—chaplain, once, to Charles I—he
had returned to England at the Restoration of
Charles II, and had been Dean of Christ Church,
Oxford, and Bishop of Worcester.

Very unlike others of his time, he had but one
meal a day, rose at five throughout the year, and
lived to be eighty-eight.[1]

[1] *Dict. of Nat. Biog.*

He contributed generously towards rebuilding
the ruins of St Paul's—founded scholarships at
Pembroke College, Oxford—gave £2200 to Christ
Church (where his portrait hangs in the great
hall)—improved Farnham Castle (country palace
of the Bishops of Winchester)—partly restored
Wolvesley Castle (near the gates of Winchester
College), and bought from Lord Cheyne, for
£4200, and presented to the see of Winchester,
Winchester House (by the Physic Garden). This
remained the London Palace of the Bishops of
Winchester until the 19th century.

His path to the river alongside the Physic
Garden remained "Bishop's Walk" until it was
built over a few years ago. The old boundary wall
is still there.

Lord Cheyne (Viscount Newhaven), who had
built Winchester House, was Lord of the Manor
of Chelsea—a good landlord, and a benefactor of
the parish. He "embellished" the Manor House,
near the Physic Garden (built by Henry VIII
for the infant Princess Elizabeth) and especially
its garden. Evelyn (1696) says: "I made my
Lord Cheyney a visit at Chelsea, and saw those
ingenious water-works invented by Mr Win-
stanley"—who built, and perished in the Eddy-
stone lighthouse in 1703.

On the north side the Garden was bounded by the road, which until lately was Paradise Row (now Royal Hospital Road).

It is difficult to give with any certainty the origin of its name. Mr Reginald Blunt in his book, *Paradise Row*, containing many interesting biographies of its residents, states that he has not been able to find any "clue to its evolution."

In the *Survey of London*, of which Mr Philip Norman is general editor, published by the London County Council, it is stated on the authority of Mr Randall Davies that the Paradise Row houses were built in 1691.

The word "paradise" is used, as everyone knows, for an enclosed pleasure garden, or park. John Parkinson, Apothecary to Charles I and author of a Latin herbal, and book on gardens, in merry mood, translates his name *Paradisus-in-sole*, "Park-in-sun."

In the 16th century Sir Thomas More made a great paradise in Chelsea. There Erasmus and Holbein were among his guests. Heywood describes it as "wonderfully charming, both from the advantages of its site, ... and also for its own beauty; it was crowned with almost perpetual verdure; it had flowering shrubs, and the branches of fruit trees interwoven in so beautiful a manner that it appeared like a tapestry woven by Nature herself." Gardens in those days were works of art.[1]

From Faulkner's map of Chelsea in 1717, and Kip's bird's-eye view of Beaufort House (once More's), in 1699, it is evident that the whole parallelogram made by Church Street and Beaufort Street on east and west, and Fulham Road and the river on north and south, was Sir Thomas More's property. Mr Randall Davies, who has done much original research in Chelsea records, in his book, *Chelsea Old Church*, goes so far as to say that the greater part of Chelsea to the west of Church Street belonged to Sir Thomas More.

In the north-east corner of this ground was the "Queen's Elm." Faulkner gives the origin of that name. Tradition has it that Queen Elizabeth, when walking with Lord Burghley, took shelter there under an elm during a storm. A seat was afterwards put round the tree, which was called

[1] The late Dr Frank Payne, in a copy of a paper read by him before the Bibliographical Society, on the early German herbals, and which he kindly gave the writer, mentions that he found, in the British Museum, the signature of Sir Thomas More in one of the copies of the *Herbarius*, printed at Mainz in 1484. The book may have been a present from Holbein or Erasmus to their hospitable host.

the "Queen's Elm"—additional evidence, if any
were wanted, that the land had once been More's;
for Burghley was at that time the owner of
More's estate, and would, no doubt, have been
showing the Queen his own land.

Later on—in 1625—this northern part was
walled in, and became "The Park"—or "Chelsea
Park"—and so it continued until a few years ago.

It is a pity that Chelsea Park was not kept
as an open space—part of a green band which
might at that time have been made round London. Fifty years ago, through its old iron gates,
which opened into the Fulham Road opposite,
and a little way beyond, the Consumption Hospital, could be seen a park of cedars, old mulberry
trees, elms and whitethorn, full of blossom in the
spring, all set in long grass—more like the
country than any London suburb.

In a letter to *The Times*,[1] an ineffectual attempt
was made to save it by the present writer. Within
a year of that date, Chelsea Park—trees and all
—had disappeared, and the bricks and mortar of
the "Elm Park Estate" were settling down upon
the "paradise" that had been Sir Thomas More's.

Now, Paradise Row was the name of a part of
the only road which led to More's house, with its

[1] November 24, 1875.

wonderful gardens and park, long after its first owner's death, *the* Paradise of Chelsea. The road passed, too, by the Bishop of Winchester's gardens, the elaborate gardens of Lord Cheyne's Manor House, and the Apothecaries' Garden, which already contained rare fruits and flowers and had European fame. It may well have been spoken of as the Paradise Road, and a row of houses built on it would thus get its name.

MAP OF CHELSEA PHYSIC GARDEN, MADE UNDER THE DIRECTION OF ISAAC RAND,
APOTHECARY, F.R.S. AND DEMONSTRATOR OF THE GARDEN, A.D. 1753

Chapter iii

So the Garden began. But all beginnings are
difficult. The Company had already learnt
that lesson. The Garden was miles away from the
eyes of the Master and Wardens at Blackfriars;
the gardener, who had £30 a year and a house, became discontented, and demanded higher wages;
workmen cheated; plants were stolen. The Great
Fire, like a great war, must have upset for a
time men's moral balance.

The Garden became an endless expense. It was
proposed that it should be abandoned. But the
majority of the Company decided on keeping it.
Private members again helped—subscriptions
came in. A high wall was built round it. Thieves
and cold winds alike were shut out.

A fresh start was made. Plants growing in a
garden belonging to a Mrs Gape at Westminster

were bought and gradually transferred to Chelsea. Rare shrubs, "nectarines of all sortes, peaches, apricockes, cherries and plums," were planted, and a water-gate was placed in the South Wall.

In 1682, within ten years of its foundation, the Garden was worthy of a visit from Dr Herman, Professor of Botany at the well-known University of Leyden. Public opinion had compelled Charles II to stop the war with Holland. Holland, like England, traded with the East, and could import rare seeds and bulbs; so an exchange was arranged. John Watts, an apothecary who had charge of the Garden, took his small cargo to Leyden, and returned with another to Chelsea.

Four cedars of Lebanon were then planted— about the year 1683. No record of the date was worth keeping. A long life for them in a foreign climate was not to be expected, and they were not imposing plants, although larger than the little cedar De Jussieu, the French botanist, nursed so carefully on a stormy voyage from Syria: planting it in his spare hat, and sharing with it his scanty allowance of only half a pint of drinking water a day. But two of them became the celebrated "Chelsea Cedars." Many Londoners will remember the two picturesque trees, standing like sentinels, one on each side of the entrance to the Garden.[1]

Two planted in the middle of the Garden were cut down in 1771 and sold as timber. They kept sunshine from the flowers. But of the two which stood on either side of the iron gates, one lived till 1878, the other till 1903.[2] Many times they must have been sketched. An engraving of them is to be seen in Johns' *Forest Trees of Britain*— the Rev. C. A. Johns, who was Kingsley's tutor, and inspired him, as his books have many others, with a love of Nature; and a small woodcut in Selby's *Forest Trees*, 1842, shows one of them with lower branches, which were afterwards broken off by snow.

These Chelsea cedars, as they grew, must have been watched with the greatest interest. No cedars had ever been seen in England, though every child had heard of the wonderful trees. Gerard's *Herball* must have been consulted. This is his description : " The great cedar of Libanus is a very big and high tree, not only exceeding all other resinous trees, but in its tallnesse and

[1] The frontispiece shows the trees as they were about a century ago.

[2] There are good photographs of the survivor in *History of Gardening in England*, by Hon. Mrs Eustace Cecil (Hon. Lady Cecil), and in *Chelsea Reach*, by Reginald Blunt

largenesse far surmounting all other trees...in shape like a sharp-pointed steeple."

Londoners must have thought that their great-grandchildren would see trees like Salisbury spire towering over houses in Paradise Row.

The average height of a well-grown cedar is said not to exceed 80 feet. Still, on their native mountains of the Lebanon, there must be a grandeur in the old cedars far surpassing that of the cedars of the Atlas. The wide-spreading flat boughs, dense and dark green, must be more impressive than the comparatively thin, pale, scattered foliage of the Algerian cedar. Some of the best trees, too, when the writer happened to see part of the Algerian Atlas years ago, had been destroyed by the natives—a protest against the occupation of their country by the French.

Other cedars of Lebanon may have been grown from some seeds procured by Evelyn, but three of the Chelsea cedars were the first cedars in England to produce cones—in 1732. They may have been the first planted.[1] From their cones, trees were raised in other gardens, and the

---

[1] The Hon. Lady Cecil, in *London Parks and Gardens*, after examining all the evidence, considers "that the Chelsea trees' claim to be the first is fairly established."

Lebanon cedar became well known throughout the country.

But there will be no more cedars in London until London smoke disappears. Cedars are choked by it. All the old cedars round London are now dead or dying. The last of those planted by Lord Holland at Holland House is scarcely alive.[1]

A young medical student, Hans Sloane, destined in days to come to play a great part, not only in the history of the Physic Garden, but in the history of Chelsea, was at this time corresponding with John Ray,[2] one of the founders of modern Natural History.

In Ray's *Philosophical Letters*, published in 1718, a letter from Dr Sloane gives an account of a visit to the Garden in 1684: "I was the other day at Chelsea, and find that the artifices us'd by Mr Watts have been very effectual for the Preservation of his Plants, insomuch that this severe winter has scarce kill'd any of his fine Plants. One thing I much wonder to see, the

---

[1] A cedar in Kensington Gardens, near the Albert Memorial, is an exception. Stunted and wizened, like a London slum child, and not much larger than the moss-covered oaks on Dartmoor, it still has a crop of bright green needles on its flat boughs every summer.

[2] Linnæus said that Ray's *Catalogue of English Plants*, 1670, introduced the "Golden Age of Botany."

Cedrus Montis Libani, the Inhabitant of a very different climate, should thrive here so well, as without Pot or Green House to be able to propagate itself by Layers this Spring. Seeds sown last Autumn, have as yet thriven very well, and are like to hold out."

Next year Evelyn paid the Garden a visit. He must have been looking at the foundations of the Military Hospital close by, with its great quadrangle, which Sir Christopher Wren was copying from Cardinal Wolsey's " *Tom Quad.*" at Oxford. Hot-houses for plants from tropical climates were being tried, and Evelyn wrote:

Aug. 7, 1685. I went to see Mr Watts, keeper of the Apothecaries' Garden of simples at Chelsea, where there is a collection of innumerable rarities of that sort; particularly, besides many rare annuals, the tree bearing Jesuit's bark, which had done such wonders in Quartan Agues—what was very ingenious was the subterranean heate,[1] conveyed by a stove under the conservatory, all vaulted with brick, so as he (John Watts) has the doors and windows open in the hardest frosts, secluding only the snow.

[1] "An arrangement more efficient than the open fire-baskets formerly in use at Oxford."—Arthur W. Hill in *Annals of Missouri Botanical Garden*, 1915. The "fire-basket" was on wheels, contained burning charcoal, and was drawn to and fro in the conservatory. The underground flue was afterwards abandoned. It dried the soil, and was dangerous.

The Garden Committee had done well. It could have been no easy task to obtain "the tree bearing Jesuit's bark."

The story of the search for Cinchona trees and seeds is as full of adventure as the search for the Golden Fleece. Fate was persistently against these trees being carried off from their lonely forests on the slopes of the Andes.

As far back as 1639 the Countess of Chinchon, wife of the Spanish Viceroy, had brought Peruvian bark to Spain, to the relief of ague-stricken labourers on her husband's estate; Jesuit missionaries had brought it to Rome, where it was much needed; Robert Talbot, an apothecary, had cured Charles II's ague with the powdered bark, and made his fortune. But it was two centuries before the living plants could be imported into countries where they would grow into trees.

Many explorers tried and failed—one was murdered. Jussieu, the French botanist, procured plants, and lost them in a storm at the mouth of the Amazon, after the long river journey. Dr Royle[1] in 1839, in 1853, and again in 1856,

---

[1] Dr John Forbes Royle, naturalist, surgeon in the service of the East India Company, and accurate observer and writer, author of *Botany and Natural History of the Himalayas*, 2 vols. quarto, 1839.

when he was dying, petitioned, in vain, the old East India Company to introduce them into India.

Meantime, Cinchona forests were diminishing, and quinine, the important constituent of "bark," which a French chemist had succeeded in extracting, remained at a price beyond the reach of the mass of mankind.

In 1852 the Dutch Government sent an expedition to bring plants to Java. Few of the young trees survived the voyage, and those of a kind which produced little quinine. But the Dutch persevered, and had the honour of being the first to establish a Cinchona plantation in the Eastern world.

It was not until 1862 that Sir Clements Markham, and those who worked with him, after adventures and dangers, all recorded in his book *Peruvian Bark*, succeeded in bringing plants and seeds to India.

That was the beginning of the Cinchona forests in the Neilgherry Hills, Sikkim, Burma and Ceylon. Quinine is now within reach of all. Overworked women in a hospital out-patient room, who beg for "another bottle of Queen Anne Medicine," can have it. It has banished ague from England, although the mosquito remains, a mischievous messenger, ever ready to carry the poison if it can find it in the blood of some returned traveller.

Those who have read Daniel Defoe's *Tour Through the Eastern Counties in* 1722, know the scourge ague once was in England—and could be again if it were not for the "tree bearing Jesuit's bark."

In India it must have saved many millions of lives. If the tree had been known to ancient Rome and Greece, it is impossible to say what the course of European history would have been; for the increasing weakness of both nations, with the desolation of places like the Pontine Marshes, was due to the increasing power of the malaria mosquito.[1]

When Evelyn was visiting the Garden, other matters were occupying the minds of the Apothecaries.

James II had carried the doctrine of the divine right of kings to its logical conclusion. He had changed the heads of colleges at Oxford and Cambridge, from Protestant to Roman Catholic.

---

[1] The want of sufficient quinine was producing disastrous results at the end of the Great War. The situation was saved by the Dutch handing over at a moderate price the entire product of the Java plantation. Martindale's *Extra Pharmacopœia*, 1924, Vol. I, p. 709.

He had also revoked the Apothecaries' Charter;
and had substituted for the old Livery the names
of those who would vote as he wished. For three
years the Apothecaries' Company was a political
machine.

But throughout the land the necessary, blood-
less revolution had begun. Even Irish troops
quartered at Hounslow, as a warning to the City,
could find no excuse for fighting, and the New
Year, 1689, saw preparations for the coronation
of William and Mary.

Henry Hyde, 2nd Earl of Clarendon, son of
the historian of the Civil War—though openly
opposed to James' extravagant despotism—remained loyal, and could not bring himself to take
the oath of allegiance to William of Orange, and
to the Queen who had stepped so lightly into her
father's place.

To ease his troubled mind, and take refuge
from the perplexities of the time, Clarendon
made his way to the quiet Chelsea Physic
Garden, sat by the growing cedars, and found
peace.

He wrote in his diary: "May 17, 1689, Friday.
Being my usual fast day, I was for above two
hours at the Apothecaries' Garden at Chelsea,
where I was not disturbed by any company."

THE PHYSIC GARDEN WITH RIVER IN DISTANCE

Red Horse-chestnut in foreground shows the uneven growth of the stem of a grafted tree

And three days later: "Towards evening I went to the Apothecaries' Garden."[1]

The Rev. Dr Hamilton describes the Garden in 1691 thus: "Chelsea Physick Garden has a great variety of plants both in and out of greenhouses; their perennial green hedges and rows of different coloured herbs are very pretty; and so are the banks set with shades of herbs in the Irish stitch way." One of the old prints of the Garden shows what was meant by "Irish stitch way."

A few years later, Bowack, in a *History of Middlesex*, wrote of Chelsea: "This happy spot is likewise blest by Nature with a peculiar kind of soil which produceth nine or ten rare physical plants not found elsewhere in England, and the Apothecaries' Garden here lying upon the Thames side is a clear instance of the opinion the learned Botanists of their Society had of the aptitude of the soil for the nourishment of the most curious plants."

In 1693 it was again proposed to abandon the Garden, and again the old botanists won the day. Samuel Doody, an apothecary in the Strand, a Fellow of a scientific society which Charles II

[1] Two years afterwards Clarendon was sent to the Tower. Extremists on both sides were increasing political troubles.

had lately founded, and called the "Royal Society," agreed to look after it.

Doody was a devoted botanist. He loved the plants, especially for their "medicinal usefulness." His work was praised by the great Ray, and by Jussieu, the French professor.

Adam Buddle (commemorated by the Buddleas), of Catharine Hall, Cambridge, who left to Sir Hans Sloane an accurate account of British Flora in manuscript, said of Doody: "He generally wanted words to express his wisdom, but when he did exert himself, his discourse was full of argument and sound reasoning....In Botany...none before him ever knew so much."

But Doody wrote practically nothing. Trimen and Dyer, in *Flora of Middlesex*,[1] say that he was "a good example of that class of scientific men who, without themselves printing, are yet of the greatest help to those who do. Such men may exercise as great an influence on the progress of science as the most prolific authors."

Soon after Doody's death in 1706, James Petiver—also a Fellow of the new "Royal Society"—became Demonstrator at the Garden. Born in Warwickshire, educated at Rugby Free

[1] Trimen and Dyer, *Flora of Middlesex*, p. 378.

School, apprenticed in London, his life became a full one.

He had a large practice as an apothecary in Aldersgate Street, and was Apothecary to St Bartholomew's Hospital, and to the Charter House. He accumulated an extraordinary natural history collection, for which Sir Hans Sloane offered £4000. He helped Ray (who speaks of him as "*mei amicissimus*") with his *History of Plants*; and he published two folio volumes on natural history, with hundreds of engravings, and yet he found time to act, for years, as Demonstrator of Botany at the Physic Garden.

There are volumes of his dried plants in the Natural History Museum in Cromwell Road.

Petiver's plates represent all branches of natural history. Everyone who looks at the old natural history books must be struck by the extraordinary contrast between the beautiful illustrations of flowers and the monstrous illustrations of beasts and birds. A great artist like Albert Dürer can make accurate water-colour drawings of his tame "Belgian" rabbit, or of a Little Owl (its worn tail-feathers showing how long it had been kept in a cage!); but the book illustrations of birds and beasts up till quite recent times are grotesque, and often imaginary.

The reasons no doubt were that (as has been already pointed out) plants were a more necessary and consequently an earlier subject of study for mankind, that they were more accessible, and that they were easier subjects for the artist. They were perfect "sitters." Not so the animals. It is only in quite recent years, owing to the growing love for natural history, that the world has seen the beautiful drawings of birds and beasts which delight the present generation of naturalists.

Midway in time, between the good drawings of plants and the good modern drawings of animals, come the good drawings of butterflies. A rare book, Harris' *The Aurelian*, 1766, with its engraved and coloured plates, was one of the works which led the way. But long before that time, accurate drawings of plants had been a delight to many who knew little of scientific botany.

In 1713 again came the Apothecaries' ever-recurring difficulty of meeting the expenses of the Garden with a small balance at the bank—a Garden, too, which was held only on a lease, so that all improvements would some day—not a distant one—become the property of the land-lord.

Lord Cheyne had offered the Apothecaries the freehold for £400—a sum beyond their means. They had not even money for the repair of the barge, so that in the Lord Mayor's pageant that year the Apothecaries' Company was represented on land only, not on the river. The barge had to be laid up, and the bargemaster's salary saved.

But the Company decided that, whatever their poverty, the Garden must be carried on "for the honour of the Society, and the benefit of its younger members." Poverty, when it does not crush, is ever an effective spur.

So the Company passed a rule that every member should be taxed—every "master on binding an apprentice"—every "apprentice at the time of binding"—everyone dining at a "herbarizing" dinner; and an extra fine was imposed on those who refused to take office. All these excess profits were to be spent on the Garden.

The Company, too, did well in getting rid of some South Sea stock "with advantage" before the bubble burst.

Then came the great event in the history of the Garden.

Sir Hans Sloane was now their landlord. He

had purchased the Manor of Chelsea ten years before,[1] and although he was living and practising as a physician in the fashionable district of Bloomsbury, he must have enjoyed the short drive to his country estate, and a walk round the Physic Garden. Petiver, the late Demonstrator of Plants, had been his friend, and at Petiver's funeral, in 1718, Sloane had been a pall-bearer.

So the Apothecaries laid their troubles before him; and 200 years ago—in February, 1722 —a deed was signed by which, for a yearly payment of £5, Sir Hans Sloane conveyed the Physic Garden with its greenhouse, stoves, and barge-houses to the Apothecaries' Society "to hold the same for ever"—and so "enable the Society to support the charge thereof, for the manifestation of the power, wisdom and glory of God in the works of Creation"; and show how "useful plants" may be distinguished from "those that are hurtful."

But Sir Hans Sloane was not an indiscriminate giver. He took care that the Garden should not remain in idle, or in inefficient hands. So a stipulation was made that every year, for

[1] From the second Lord Cheyne, who had left Chelsea to become Lord Lieutenant of Buckinghamshire.

forty years, fifty specimens of plants (all from the Garden, and no two alike), carefully dried, mounted, and named, should be sent to the Royal Society. This ensured that 2000 different species of trees, shrubs and flowers would be grown in the Garden during that time. The agreement was faithfully kept.[1]

There was another condition. If the Garden was not kept up as a scientific Garden it was to be offered to the Royal Society, and, if the Royal Society refused to take it, to the College of Physicians under the same conditions, and if the physicians refused, it was to revert to Sir Hans Sloane's heirs.

That, too, was a wise precaution. It kept any future generation of Apothecaries, in want of funds to meet some unexpected expense, from the temptation to sell such a valuable building site.

[1] More than 3000 were sent to the Royal Society; they are now preserved in the Natural History Museum.

Chapter iv

IT is evident that it is to Sir Hans Sloane, beyond all others, that the Physic Garden owes its existence at the present day.

Hans Sloane was born in 1660, in County Down, Ireland, where his father—a Scotchman— was Receiver-General of Taxes. His mother was a daughter of Dr Hickes, Prebendary of Winchester. Hans, the youngest of seven brothers, was an intelligent child, devoted to natural history; but, at sixteen, all studies were cut short by an attack of hæmorrhage from the lungs—due no doubt to tubercle, "consumption" being at that time the prevalent plague in Ireland.

His illness was a blessing in disguise. Paradoxical as it seems, his long and successful life

SIR HANS SLOANE (1660–1753) PRESIDENT OF THE
COLLEGE OF PHYSICIANS, AND OF THE ROYAL
SOCIETY—LORD OF THE MANOR OF CHELSEA.
DONOR OF FREEHOLD OF PHYSIC GARDEN TO
SOCIETY OF APOTHECARIES (1722)

may have been partly due to this early attack of a dangerous malady. He was laid up for three years—the drawn sword over his head often painfully visible. During that time he learnt that, in order to prevent the recurrence of those alarming attacks, he had to lead a most careful and temperate life, and become almost a teetotaler. He must have learnt, too, perhaps unconsciously, numberless unforgettable lessons in the treatment of his own illness, which would help to make him the successful practitioner he afterwards became, in treating the illnesses of others.

An almost exact parallel to Sir Hans Sloane's history is that of a well-known physician of Victorian times,[1] who, after a like illness had subsided, became a wise and deservedly popular physician, capable, like Sloane, of hard work, helpful to crowds of patients, and eventually a successor of Sloane as President of the College of Physicians.

As soon as he was strong enough, Sloane came to England, studied physic at the Apothecaries' Hall, and botany at the Chelsea Garden.

A youth of parts, he was attracted to minds like his own—John Ray, the great naturalist, Robert Boyle, the founder of scientific chemistry,

[1] The late Sir Andrew Clark.

who had been offered, and had refused, the
Presidency of the Royal Society and a peerage,
and Dr Sydenham, the preacher of common-sense
in medicine, were his friends.

After four years of study in London, Sloane
attended lectures in Paris, and at Montpellier,
where there was a botanic garden; and, while in
France, was advised to take a medical degree at
Orange-Nassau, which he did "with applause."

In 1684 he returned to London, sent an account
of his travels to Boyle, a collection of rare plants
to Ray, and went off to see what had happened
to the cedars in the Physic Garden.

The next year the Royal Society elected him
a Fellow, and in 1687 the Royal College of
Physicians did him the same honour.

He was then asked to accompany the Duke of
Albemarle, who had been made Governor of
Jamaica. The journey took more than three
months; but to anyone who cared for the ever-
changing face of sky and sea, the smokeless air,
the strange birds and fishes, a voyage in a great
sailing ship must have had a charm, of which
modern travellers dashing through the water
high up on the deck of a great steamship know
nothing.

The Duke of Albemarle fell ill and died; and

after fifteen months in Jamaica, Sloane accompanied the widowed Duchess to England. It must
have been a depressing time. Happily for Sloane
he was a naturalist; and naturalists are not dull
in such a country. Sloane collected plants, of
which he brought back 800 specimens, and made
notes for his future voluminous history of the
island, its flora and fauna.

Some of the living fauna did not survive the
voyage. An iguana jumped overboard. A crocodile
found its way into a tub of salt water and died.
A yellow snake, seven feet long, escaped, lived
on the roof of the deckhouse, and fed on the
ship's rats; until some passengers who were not
naturalists—"footmen and other domestics of
her Grace, being afraid to lie down in such company, shot my snake dead." They, no doubt,
agreed that Æsculapius should keep his serpent
to himself!

There are no details of the death of the snake.
The execution may have been carried out when
Sloane and the captain were safely ashore. It is
impossible, too, not to suspect the Duchess of
being on the side of the conspirators. But it seems
a noisy way of getting rid of a harmless snake,
which could have been reached with a boat-hook
and thrown overboard—and certainly damaging

to the ship's furniture. But the footmen probably refused to touch it, even at the end of a boat-hook.

This dread of snakes seems universal—not only in man. At the Zoological Gardens some time ago when a snake was being carried through the ape-house, the chimpanzees, which must have left Africa mere babies, knowing little of snakes, were terror-stricken—their hair stood on end—they clung to the top of their cage even after the snake had been removed, and were ill for some time afterwards.

Sloane and his collection arrived safely in England, and Evelyn reported on it in his diary:

April 16, 1691. I went to see Dr Sloane's curiosities, being an universal collection of the natural productions of Jamaica, consisting of plants, fruits, corals, minerals, stones, earth, shells, animals and insects, collected with great judgment; several folios of dried plants, and one which had about 80 sorts of ferns, and another of grasses; the Jamaica pepper in branch, leaves, flower, fruit, etc. This collection, with his journal... very copious and extraordinary, sufficient to furnish a history of that Island.

A history of the island was published by Sloane some years afterwards—two large volumes with nearly 300 engraved plates. It had been preceded by a *Catalogus Plantarum* in 1696.

Sloane settled in London, took a house in

Great Russell Street, near the very spot to which, after his death in 1753, his great collections were moved. He became physician to Christ's Hospital (to which he returned the salary he received), and secretary of the Royal Society, of which he revived the "Transactions." He married Mrs Rose, widow of a wealthy Jamaica magnate, took an M.D. degree at Oxford, received honours from several foreign Academies, and was appointed Physician to Queen Anne.

In 1716, George I made him a baronet—the first physician to receive that honour. In 1719 he was elected President of the College of Physicians, and held the office for fifteen years. Among other benefactions he gave the College a donation of £100. Sir Isaac Newton dying in 1727, Sir Hans Sloane succeeded him as President of the Royal Society, retiring after fourteen years, to the regret of the Committee.

In 1712, Sir Hans Sloane purchased the Manor of Chelsea—and with it the Physic Garden, which had taught him botany, and which he presented to the Apothecaries in 1722—but he continued to live and practise in Bloomsbury. It was not until 1742 that he moved with his vast collections to the Manor House in Chelsea.

His practice became large and fashionable, but every morning until ten o'clock he saw patients gratuitously. He was a generous supporter of many hospitals. He believed in the virtue of Peruvian bark, and invested the money he received for the Jamaica expedition in purchasing it, and has the credit of making that important medicine popular; but he was cautious in the use of drugs, and, like Sydenham, seems to have been successful chiefly from his own power of observation at the bedside.

Sloane has been ridiculed for using in his practice an ointment of viper's fat. But that must have been only a small part of his treatment.

Every generation, too, must inherit—in other matters than in medicine—imperfect truths from earlier generations. These well-worn clothes it seems sometimes better to mend, and use, than hurriedly to throw aside. We may feel chilly without them!

The virtue of viper's fat is still firmly believed in, and the fat is collected and used in many country districts. "Brusher Mills," a Caliban of modern days, who, a few years ago, lived under the trees in the New Forest, and caught snakes for the Zoological Gardens, found a ready sale

MARBLE STATUE OF SIR HANS SLOANE, M.D., P.R.S.

Plant of medicinal rhubarb in foreground

for the fat of vipers killed in the autumn.
Whether the fat of vipers has, or has not, a
greater virtue than that possessed by the fat
of other animals, it would be difficult to say.
Happily there is not enough of it to allow of
any extensive trial.

During his last years at the Chelsea Manor
House Sloane felt, as old people do, the loneliness
of life. He must have made many friends older
than himself—men from whom he could learn.
They had all gone. John Ray, half a century
before, had written a touching farewell letter to
his "best of friends." Even his contemporaries
had gone. Like Tithonus, "cruel immortality"
seemed to oppress him.

His treasure house, with its wonders of Nature,
and man's art, was some consolation; and George
Edwards (the librarian at the College of Physicians—a natural history artist, and author of
the best book on birds of its date) came once a
week to see him and bring news. Then in 1753
he suddenly dropped (as he said he would), "like
a ripe fruit," and the consumptive boy passed
away just before reaching his 93rd birthday.

His great and varied collection, which he
valued at £80,000 (the gold and silver coins and
medals were worth £7000 as bullion), he left to

the nation on condition that £20,000 was paid
to his daughters, Lady Cadogan and Mrs Stanley.

Horace Walpole—a little out of his element
as one of the many trustees of the will—wrote
to Sir Horace Mann:

Feb. 14, 1753. You will scarce guess how I employ
my time; chiefly at present in the guardianship of
embryos and cockleshells. Sir Hans Sloane is dead,
and has made me one of the trustees to his museum,
which is offered for twenty thousand pounds to the
King (or) Parliament (or on refusal) to the Royal
Academies of Petersburg, Berlin, Paris and Madrid.
He valued it at fourscore thousand, and so would any-
body who loves hippopotamuses, sharks with one ear,
and spiders big as geese! You may believe that those
who think money the most valuable of all curiosities
will not be the purchasers. The King has excused
himself, saying that he did not believe that there are
twenty thousand pounds in the Treasury. We are a
charming wise set, all philosophers, botanists, anti-
quarians and mathematicians; and adjourned our first
meeting, because Lord Macclesfield, our chairman, was
engaged to a party for finding out the longitude.

After much discussion, Parliament decided
to accept the bequest—raise the money by a
lottery—move the 50,000 books and manuscripts, 23,000 coins and medals, 3000 gems
and antiquities, 16,000 objects of natural history

and anatomy, to Montagu House, and form a British Museum.

The monument to Sir Hans Sloane, with its large urn and serpents, in the graveyard of Old Chelsea Church, was erected in 1763.

It has been suggested that as Sloane was for many years an absentee landlord, and pulled down Beaufort House, it would have been better for Chelsea if he had "never been born." But, without any doubt, Beaufort House, which had a frontage of 200 feet, and had been for twenty years unfurnished and unoccupied, would have disappeared as soon as it became an eligible building site—whoever happened to be the owner.

Chelsea, too, is greatly indebted to Sloane for saving the Physic Garden. It was also due to his influence in high places that an order was sent from the Lords of the Treasury to the King's Surveyor to open the King's Road to Chelsea residents. The King's Road was originally a narrow cartway. It allowed farm labourers and market gardeners access to their fields on each side, and to Chelsea Common on the north. It was widened by taking land off the headlands, where ploughs turned in the fields by the side of it. As a compensation the tenants were allowed to make use of the road.

In Sloane's time the King's Surveyor took upon himself to close the gates on the road against all but a few privileged persons. Sloane, supported by three freeholders, after some trouble, obtained an order to Brigadier Watkins, Surveyor of the King's private roads, to allow the tenants of the fields "free passage with their carts and horses"—and also to open again the "ditches lately filled up."

The ditches must have not only drained the fields, but prevented cattle from straying.

Residents in Chelsea thus had the advantage of using the King's private road without contributing to its maintenance, for it was not until 1830 that the parishes, through which it passed, became responsible for its repair.

That Sloane was liked by his patients, rich and poor, is evident. That he was popular with fellow scientists is shown by the way in which Fellows of the Royal Society took his part in a quarrel.

He could not have become rich and successful without running the gauntlet of criticism. At Royal Society meetings Dr Woodward thought that Sloane gave himself airs. Woodward used to scowl at him across the table, and finally

made insulting remarks when Sloane was reading
a paper.

The Royal Society, under the presidency of
Sir Isaac Newton, took Sloane's side, voted for
Woodward's expulsion, and refused to reinstate
him. Woodward was a notoriously pugnacious
person. He fell out with Dr Mead, George II's
physician. The quarrel, begun with words, was
continued with walking-sticks, and ended with
swords.

Woodward slipped and fell. Mead offered him
his life if he would beg for it. Woodward is
reported to have said that he would take Mead's
offer, but not his physic. Possibly the sentence
was not finished until Woodward was on his
feet again, and swords were safely sheathed.

The high opinion of Sloane held by foreign
men of science is shown by Kalm's diary.[1]

Edmund Howard, Sloane's literary caretaker
in the empty Beaufort House, said of his master
that although he "had been acquainted with men
superior to him both in natural talents and
acquired accomplishments," he was "easy of
access, very affable, and free in conversing with
all who had any concerns with him, and a good

[1] See p. 86 *infra*.

master to his servants, for they lived many years
with him. He was also a good landlord, and
never, that I know or heard of, did one harsh
thing by any of his tenants."

Pulteney says that he was engaging in his
manners, and obliging to all—that he kept an
open table once a week for his learned friends—
that he was a Governor of almost every hospital[1]
in London—that to each he gave £100 in his
lifetime, and left a more considerable legacy at
his death.[2]

Of Sloane it may fairly be said that he served
his generation.[3]

Many years after his death, when the descendants of Sloane in Chelsea, and the descendants
of Mary Davies in Mayfair, had allowed streets
and squares to cover the fields of the Manor of
Chelsea and the damp meadows of the Manor
of Ebury—when Hans Town and Belgravia had
joined hands over the Westbourne—a street had

[1] Linnæus, lecturing at Upsala in 1741, said: "The hospitals
of London, both for number and goodness, excel all others."

[2] Dr Richard Pulteney, *Progress of Botany*, 1790, Vol. II,
p. 84.

[3] The conventional engraving of Sloane probably conveys as
little of the character of the man as the bust of Shakespeare, with
its wooden expression, or the engraving in the First Folio, do of
the poet.

been made on the eastern boundary of Chelsea,
and given the name of the man who 200 years
ago saved the Physic Garden, and by his Will
founded the British Museum. It was an inspira-
tion; for Sloane Street well represents the life
of Sir Hans Sloane. Those who walk all the way
down it know that it is very *long*, obviously
prosperous, and perfectly *straight*!

Chapter v

So the year 1722 brought new life. No prescription of Dr Sloane's could have restored a sick patient as his wise gift did the fading fortunes of the Physic Garden.

A Garden Committee, including Master and Wardens, was at once formed. James Sherard was among its members—an apothecary in Mark Lane, a well-known botanist, Fellow of the Royal Society, friend of Ray and Petiver, and brother of William Sherard, the Fellow of St John's College who founded the Sherardian Professorship of Botany at Oxford.

James Sherard had been apprenticed as

THE PHYSIC GARDEN IN SPRING
Showing entrance from the Embankment

Apothecaries' assistant to John Watts, a former manager of the Garden, and had learnt his botany well. A description of James Sherard's garden at Eltham, with its rare plants, in two folio volumes with 300 plates, was written by Dillenius[1] of Oxford in 1732.

The new Committee decided to discharge the gardener, paying him a quarter's salary, and to appoint, probably at Sir Hans Sloane's suggestion, Philip Miller—an able man, and well trained in practical gardening by his father, a nurseryman.

Two years at the Garden enabled Miller to publish *The Gardener's Dictionary*—a great book, destined to go through many editions, to be translated into other languages, and to give its author the right to be styled, even by foreign botanists, the chief of gardeners—*Hortulanorum princeps*. Linnæus called the book a "dictionary, not only of horticulture, but of botany." Later on Miller published two folio volumes with 300 plates drawn from plants in the Physic Garden. Miller was the first to notice the part played by

---

[1] Dillenius was the first Professor of Botany at Oxford. "He was in the habit of scattering seeds in the neighbourhood of the city, some of whose descendants caused surprise to later generations of botanical students."—Vernon's *History of the Oxford Museum*.

insects in fertilizing flowers, and became an
F.R.S.

The Committee also appointed Isaac Rand (an
apothecary, F.R.S., and a zealous botanist)
Director of the Garden and Demonstrator of
Plants. Rand must have found his high office
eclipsed by the work of the energetic gardener,
whose papers were continually appearing before
the scientific world. Miller, too, had not only
published his *Gardener's Dictionary*, but, without medical training, a catalogue of the medicinal
plants in the Garden. Isaac Rand was angry at
this encroachment on his province, and published,
and presented to the Apothecaries, a fuller catalogue in Latin.

It may have been this trace of hot temper
which made Linnæus give the name "Randia"
to a genus of tropical plants. It was Linnæus'
way!

Rand published the map of the Garden on
page 31.

Money was now freely spent by the Apothecaries on a Garden they owned. A yearly tax of
six shillings a head was levied on all members
of the Society; and forty pounds were given from
corporate funds—willingly, no doubt, for the
Apothecaries had just sold every penny of their

South Sea stock. Sir Hans Sloane gave a hundred pounds towards the repair of the river stairs, and it was probably at his suggestion that the College of Physicians, of which he was President, contributed an equal amount. Sir Hans was a peacemaker, for the College of Physicians, the Surgeons' Company, and the Apothecaries' Company were not always in harmony.

The Garden wharf was now thought unworthy of the new landlords, so a thousand pounds were borrowed to build a larger one. Then Miller, keen gardener, found that he could not do justice to the new tropical plants without more "glass."

Two hot-houses and a greenhouse were built for him, and finished in 1732, and Sir Hans Sloane came from Bloomsbury to lay the foundation stones—as the inscription on the present greenhouse relates.

Then came the reckoning—one thousand eight hundred pounds—an unexpected bill—to be again met by generous subscribers and the Society's Corporate Funds.

Nothing daunted, the Apothecaries agreed to contribute twenty pounds a year towards the salary of a plant collector who would explore an American colony, which the warm-hearted General Oglethorpe had founded for poor debtors, and called, after George II, who had given him a charter, "Georgia."

Cotton seed was sent to the new colony by Philip Miller, in 1732, and became "the parent stock of upland cotton"[1]—the ordinary cotton of the United States. So from this little packet of seed, sent as a present from the Curator of the Chelsea Garden to the young colony in which Sir Hans Sloane and the Apothecaries were interested, the greater part of the world's cotton is descended!

It was then decided that the Garden must have a statue of Sir Hans Sloane, and that Michael Rysbrack, the sculptor who had finished a great monument to Sir Isaac Newton for Westminster Abbey, must undertake it — "Ricebank" the clerk at Apothecaries' Hall called him[2].

Two hundred and eighty pounds were voted for it, and it was finished in 1737. A worthy statue it was—and still is, owing to the fact that it was for some time kept under shelter in the greenhouse,[3] and that when it was removed

---

[1] A. W. Hill, *Annals of Missouri Botanical Garden*, p. 222.

[2] Mrs Esdaile has lately found the terra-cotta model for Sloane's statue in the British Museum.

[3] Field states that the statue was originally "in front of the greenhouse." Kalm saw it "in one room of the Orange House." It was removed to the middle of the Garden in 1748.

CHELSEA PHYSIC GARDEN

Statue of Sir Hans Sloane. Rockery in foreground

to the Garden it was covered with a sail cloth[1]
in bad weather; but the acid-laden rain of London
has already destroyed more than its original
smooth surface; Sloane's full features are becoming thin, and the President of the College of
Physicians' gown is losing its ornaments.[2]

So the Garden prospered, and obtained European fame, both for its rare plants and skilful
management.

Dr Linnæus heard of it, and decided to visit
England and see it, and its fellow Physic Garden
by the Cherwell at Oxford.

Linnæus had slowly made his way in his own
country, and had served as Assistant to the Professor of Botany at Upsala. With extraordinary
industry, and with the gift of simple and terse
language, he was engaged in classifying the whole
living world, from buffaloes to buttercups.

In botany, gathering up the threads of work
left by Ray and others (who found difficulty, in
a crowd of synonyms, of deciding what plants
belonged to what names), and keeping as far as
he could to the old classical names, Linnæus gave
every plant a name consisting of two words instead

---

[1] Barrett, p. 139.
[2] Expert opinion is being taken by the Garden Committee on
the best means of preserving this fine statue.

of a long descriptive sentence. The first word
was the surname of its family,[1] the second word
indicated the species, and he classified all flowering plants according to resemblances in their
stamens and carpels.

He knew that his work was not final, but he
brought order out of chaos, and made a great
index to the vegetable world, with names so well
chosen that most of them are in use at the present
day.

But the old botanists were not ready for the
new teaching. Linnæus came to London in 1736
with a cordial letter of introduction to Sir Hans
Sloane from Dr Boerhaave, a well-known physician and botanist. The letter stated that Sir Hans
was the only one worthy of an introduction to
Linnæus, and Linnæus the only one worthy of
an introduction to Sir Hans—"two men whose
like cannot be found in the world." The pretty
speech fell flat—Sir Hans was bored by
Linnæus.

Linnæus then went on to Chelsea, and saw
Miller at the Physic Garden, who received the
revolutionary Swedish botanist in much the same
way—thought him conceited, and ignorant—

---

[1] Family is here used for the Latin word *genus*, as it is in
Johns' universally-read *Flowers of the Field*.

especially of botany. That passed; the two men recognized that they were brothers. Linnæus was allowed the run of the Physic Garden, and afterwards wrote in his diary: "Miller of Chelsea permitted me to collect many plants in the Garden, and gave me several dried specimens collected in South America....The English are the most generous people on earth."

It is curious how seldom Linnæus' friends were made at first sight. Dillenius, the Professor of Botany at Oxford, disappointed at first, later on offered Linnæus half his salary if he would stay with him, and help him. In his own country it was at first difficult to get any recognition, but later on the students attending his lectures increased from five hundred to fifteen hundred.

A letter from Gronovius to Miller, kindly sent to the writer by the Rev. W. E. Layton (a great-grandson of Miller's), after the publication of the first edition of this book, shows that it was the Chelsea Physic Garden which brought Linnæus to England.

Gronovius, botanist, scholar, and Professor of Latin at Leyden, befriended Linnæus—read his proofs—arranged for the publication of Linnæus' *Systema Naturæ* at his own expense—and named the graceful little flower which travellers in Norway see growing under the fir trees, *Linnea borealis.*

Of the other three names mentioned in the letter, G. Clifford was a rich Dutch banker at Haarlem—a director of the Dutch East India Company (then at the height of its prosperity) —the owner of a private zoo, and of a magnificent garden which Linnæus was arranging; Dr Richardson, owner of a great garden—helper of poorer botanists, and writer of letters now among the Sloane MS.; Peter Collinson, a Quaker—partner in a firm of mercers—keen naturalist—and importer of the Ailanthus (brought from China in 1751), whose smooth, clean stem and branches, and graceful ash-like leaves, are always a pleasant contrast to the London Plane, in London Squares.

Collinson helped Lord Holland to lay out the gardens at Holland House. His own garden at Mill Hill still has trees he planted.

Gronovius' letter is in Latin, as were all letters between educated people of different nationalities at that time; and there seems to have been more friendly intercourse, and less jealousy then than now. Nations had not grown to their present size, and exclusiveness.

The following is a translation:

To the polite and learned man, Mr Philip Miller, Director of the Chelsey Garden[1] and distinguished Botanist.

Joh: Fred: Gronovius sends greetings:

You will now learn, most polite Miller, that *your earnest wish,* which you more than once expressed to me in letters last year, *that you might see the celebrated Dr Linnæus, is satisfied.* It is he himself who will deliver this my letter into your hands.

So when, *seized with the desire of seeing*[2] *the plants of your most celebrated ·garden, he comes to England,* I beg your help and favour.

Nor am I alone in asking this, for indeed the most noble Mr Clifford, the chief patron of botanists, pleads the same cause. In turn we both offer you our readiest offices.

The second copy of the *Systema Naturæ* which you wanted for the use of your (Chelsea) Library[3] has, I hope, been delivered to you, as on May 5th I handed it to Dr Richardson—unless it has been left with Mr Collinson.

Farewell, most polite Sir, and count the most celebrated Linnæus and me among your friends.

Lugdoni Bataviorum,[4] 19 Jul: 1736.

[1] "Horti Chelseyani praefecto."—Rand was "Horti praefectus." Miller was Curator.

[2] "Videndi amore captus."

[3] This copy is no longer in the Apothecaries' Library.

[4] Leyden, Holland.

So Linnæus came to the Physic Garden, a welcome guest. But a frost fell on the meeting.

One reason for these cool receptions of Linnæus must have been a curious flaw in character. Like all others, Linnæus had faults. The rule that a man—however great—out of consideration for the less fortunate, and remembering his own failures, must not boast, was ruthlessly neglected.

That Linnæus deserved the praise he bestowed on himself did not prevent the unfavourable impression he made.

The effect must have been that produced by Joseph in the old story, when, admired by parents, and conscious of power, he informed his brethren that he had been dreaming that his father and all of them would, some day, have to bow their heads to him. Those hours in the pit must have been an invaluable lesson.

Linnæus was never thrown into a pit, nor had he been at an English public school, so the habit remained through life. It was only when he was recognized as the giant he was, that the first ill impression was forgotten, and men came from all countries to learn of him.

Linnæus' description of the interview with Miller, as he told it to a friend,[1] is interesting.

—————
[1] Fée.

He says: "When I paid Philip Miller a visit, the principal object of my journey, he shewed me the garden at Chelsea, and named me the plants in the nomenclature then in use...I held my tongue, which made him declare next day, 'That botanist of Clifford's does not know a single plant.' I heard this.[1] I said to him, as he was going to use the same names, 'Do not call these plants thus; we have shorter and surer names.'

"Then he was angry and looked cross. I wished to have some plants for Clifford's garden, but when I came back to Miller's he was in London. He returned in the evening. His ill-humour had passed off. He promised to give me all I asked for. He kept his word, and I left for Oxford having sent a fine parcel to Clifford."

It would be still more interesting to have Miller's account of the interview. There must have been something more than a discussion on names to disappoint Dillenius, bore Sloane, and make Miller angry with the botanist he admired, and invited to Chelsea.

Linnæus enjoyed his visit, saw Ranelagh Gardens, and no doubt walked over the wooden

[1] Probably said in English to a bystander.

bridge at Fulham[1]—probably with Miller as his
guide—to Putney Heath, and saw a mass of
gorse.

Doubt has been thrown on the well-known
legend that Linnæus fell on his knees on seeing
it. Dr B. D. Jackson[2] says: "As the gorse is a
spring-flowering plant, and as Linnæus' visit to
England was in the late summer the tradition is
unfounded." The legend requires investigation.

First of all as to the *date* of his visit to
England—Mrs Caddy[3] writes: "The first time
Linnæus crossed Putney Heath the sight of the
gorse blossom in its *blaze of May* made him fall
on his knees in rapture."

Here the date is now seen to be incorrect.

Gronovius' letter[4] to Miller, brought by Linnæus to England, is dated July 19, and Linnæus
reached London about July 27.

The date of his departure is not given in biographies, but he stayed beyond the two months
arranged with Clifford. He was in Chelsea, within
reach of Putney Heath, at the beginning, and
end of his visit.

---

[1] Built seven years before, in 1729.
[2] B. D. Jackson, *Linnæus*, 1923, p. 158.
[3] Florence Caddy, *Through the Fields with Linnæus*, 1887,
p. 329.
[4] See p. 71.

As to the origin of the tradition, Sir James Smith—in early life contemporary with Linnæus —writes (in Rees' *Cyclopædia*, 1819) of Linnæus' "pure and ardent love of nature for its own sake," and of the tradition that the "golden bloom of the furze on...Putney Heath delighted him so much that he fell on his knees in a rapture at the sight"—Smith says that Linnæus "was always an admirer of this plant, and laboured in vain to preserve it through the winter in his greenhouse." He had seen it in bloom during a tour in Germany.

As far back as 1800 Smith had written in his first edition of *Sowerby's English Botany*, begun in 1790 (with illustrations by Sowerby), "Nor can anything be conceived more brilliant than Putney Heath in the month of May, when its [the gorse's] honey-scented blossoms spread in one sheet of gold over the country. Linnæus is reported to have fallen on his knees in a rapture of admiration at the spectacle." In the second edition,[1] published after Smith's death, the tradition is repeated, and the editor is careful to add: "Although May is the time of abundant flowering the blossoms expand the whole year if the weather be mild."

[1] 1839, Vol. VI, p. 5.

Long after Sir James Smith's time—in 1864—
an "entirely revised edition" of *Sowerby's English
Botany* was published—in twelve large volumes.
The editor states "...it is related that when he
(Linnæus) saw it in flower on Hounslow Heath he
fell on his knees and thanked God for having
created a plant so beautiful. The same story is
told of Dillenius, so that we may perhaps doubt
its authenticity."

But that is no reason for doing so. Evidently
someone had confused the name *Linnæus* with
that of the Oxford Professor of Botany, *Dillenius*,
and Putney Heath with that of the better-known
Hounslow Heath.

The other reason given for doubting the truth
of the legend requires more consideration. Winter,
spring, and early summer are, no doubt, the flowering time of Common Furze. In August flowering
is over, and the plant is in seed.

But, in favourable seasons, blossoms are to be
seen at other times—especially when the plants
have been nibbled or cut, and regular flowering
sometimes begins in the autumn.

Dwarf Furze too—a companion of Common
Furze, and by some considered a mere variety—
is in flower all the summer and autumn.

But even if Linnæus' visit was made in August,

when the seed-pods were crackling, and scattering
their seed in the sun, it must have been an impressive sight—especially to one who had tried
in vain to keep single specimens alive in a greenhouse.

Linnæus must have seen the dense, rounded
green masses of furze stretching away from the
river, over Barnes Common and Putney Heath,
into the far distance on Wimbledon Common,
dotted with patches of blossom—and have realized
the golden carpet of May. He was impulsive, and
his "most noteworthy trait," Jackson says,[1]
"was his ardent piety." So there seems to be
hardly sufficient reason for doubting a story which
was current within twenty-two years of Linnæus'
death, and, probably, during his lifetime.

Linnæus could not have seen Putney Heath
in its brilliant May dress, but a soldier recognizes
and salutes his colonel in mufti.

Linnæus spent weeks at Oxford—botanized,
no doubt with Dillenius in Bagley Wood, and
Headington Marsh—wild land even fifty years
ago—where Columbine of two colours grew, and
Grasshopper Warblers—elusive birds—nested
every summer—came back to Chelsea down the
river—and, overwhelmed by the vastness of

[1] B. D. Jackson, *Linnæus*, p. 158.

London, sailed on to Holland, when the west
wind blew.

On returning to Haarlem, Linnæus completed
the sumptuous *Hortus Cliffortianus*, and other
works—500 pages folio, and 1350 pages octavo—
before the end of the year 1737—then returned,
tired and homesick, to Sweden, where the well-
known honours awaited him.

The end of his life was not happy. Madame
Linné is described[1] as a "rough matron with
coarse features"; but she was a thrifty house-
keeper. Her daughters were up and spinning at
four o'clock, and ears were boxed if the spinsters
were late.

Fabricius, Linnæus' devoted pupil, said:[2] "She
often drove pleasure from our society." Madame
Linné must have spoilt much interesting talk.
Good conversation, as many know, can be checked
by a word out of season—as the sensitive plant
droops at a breath; and is slow to revive.

In 1774 Linnæus' appearance changed. A
"stroke" paralysed mind and body—memory
went—thoughts became muddled, and speech
incoherent. He should have been allowed to rest
out of sight of the world. The great weather-

---

[1] B. D. Jackson, *Linnæus*, p. 312.
[2] *Ibid.* p. 313.

beaten ship, which had brought home such treasures, should have found a quiet harbour.

But Madame Linné, who ruled the house,
urged the broken Linnæus, for the sake of small
fees, to continue his ineffective examination of
students, and lectures no one could understand[1]
—a pitiable sight. Then death came to the
rescue, and in January 1778, Linnæus passed
away.

Among notes of a like kind, Linnæus records
in his autobiography that he "read the earth,
minerals, vegetables, and animals as in a book"
—that he was "one of the greatest observers we
have had"—that he was "permitted to see more
of creation than any mortal before him"—that
"no one performed the duties of his professorship
with greater zeal"—that "no one was a greater
botanist or zoologist," or "became so celebrated
all over the world."—All true.

And he is light-heartedly critical in some of
the names he gives to plants. One, for instance, he calls *Buffonia tenuifolia*—the thinly-
leaved Buffonia—to commemorate the French
naturalist's scanty botanical knowledge.

But it was an age of brilliant discoveries, and

[1] B. D. Jackson, *Linnæus*, p. 337.

Linnæus was the guiding spirit. He seems even
to have had a vision of that great drama of the
succession of life on the earth; at which we all
now sit, at no cost to ourselves, wondering spec-
tators.

He said that he had "never seen the fragments
(*rudera*[1]) of a universal deluge," but "*successiva*"
rudera; that he "would willingly believe the
earth to be older than the Chinese assert had
the Scriptures allowed him"[2]—and he wrote in
a preface to his museum catalogue: "The innumerable petrifactions...of animals never seen by
any mortal in our days which lie hid among the
stones under the most lofty mountains...reach
far beyond the memory of any history."

It must have been an awe-inspiring discovery
—a revolution in thought difficult for the present
generation to realize. Dr Garden, of Carolina,[3]
wrote: "It is to Ellis[4] and to Linnæus I owe the
placing of me in a land of wonders."

For a time his books were burnt at Rome.

So Linnæus had to be careful of his words.
His friend Fabricius wrote: "He dared not pub-

[1] Literally, builder's rubble.
[2] Pulteney's *Linnæus*, 2nd edit., 1805, p. 365.
[3] and of the gardenias.
[4] John Ellis, a naturalist and geologist living in the West
Indies.

lish many important observations relating to the general arrangement of nature because he was afraid of the excessive violence of the Swedish divines...who, too faithful to their own arguments, do not consider that nature, as well as revelation, proclaims...the hand of that great master who formed both."

But we must not rail at those "divines." The new fermenting wine was a danger to the old bottles; and they must have our sympathy in their dread of being hurt by it.

From childhood the very letter of Genesis had been to them an inspired statement of scientific fact. They were ready to die for it. To doubt it would, they felt, throw doubt upon all they trusted as guidance of life.

It must be remembered, too, that although "divines" get the credit for opposing the advancement of learning, it is often the over-zealous layman who makes the fiercest attacks on new knowledge.

Chapter vi

Peter Kalm, pupil of Linnæus, visits Garden in 1748; walks in footsteps of Linnæus to Putney Heath to see yellow Furze; describes the greenhouses; visits Sir Hans Sloane and Museum; considers Chelsea Garden a rival of Gardens of Paris and Leyden. —News of Kalm's arrival in Sweden cures Linnæus of gout.— Linnæus' pupil, Fabricius, arrives in Edinburgh, 1767; rides to London, collecting plants and insects on the way; sees Banks off on Cook's first voyage.—Fabricius again in England in 1775 and 1787; visits Chelsea, and describes 200 new species of butterflies from the drawings of William Jones in Manor Street; becomes the "founder of scientific entomology."—William Hudson appointed Demonstrator.—Philip Miller pensioned.—Two Cedars cut down in 1771.

L INNÆUS' inspiring genius attracted devoted pupils. These he sent into all countries to study natural history and bring back specimens to Upsala. Two of them came to Chelsea.

In 1748 Peter Kalm, a Swede, and lecturer on economic botany, landed in England.

Here he was detained six months, "for want of a vessel," he said, "to cross to America." The delay produced an interesting diary, with voluminous notes on horticulture, agriculture, and the customs of the English. It was translated by Joseph Lucas in 1892.

Kalm paid many visits to the Physic Garden.

On April 22, 1748, he writes: "We saw Chelsea *Hortum Botanicum*, which is one of the principal ones in Europe. Here we found the learned Dr Miller, who is *Horti Præfectus* of the same. In the evening I was at the house of Dr Mortimer, Secretary of the Royal Society. Here I met the great Ornithologus, Mr Edwards, who had published a book on birds in the English language, with matchless copperplates, all in lifelike colour, so that it looked as if the bird stood living on the paper."[1]

On May 10th Kalm remarks that "the land round Chelsea is almost entirely devoted to nursery and vegetable gardens. The same is true

[1] George Edwards was the librarian to the College of Physicians who paid Sloane weekly visits at the Manor House at Chelsea. There is a good copy of his work on natural history—seven folio volumes full of excellent plates, chiefly of birds, etched on copper and coloured by hand—in the library of the College of Physicians, presented by its author. The volumes appeared between the years 1743 and 1746, and were published by the College of Physicians—the only work the College ever issued.

It was afterwards translated into French, German and Dutch. Edwards possessed an authentic painting of the Dodo, taken from life, now in the British Museum.

Gilbert White began the notes on natural history, which led to the *Natural History of Selborne*, just after the publication of Edwards' volumes. It would be interesting to know whether White had access to them through either of his two brothers, Benjamin White, the natural history publisher, or Thomas White, the writer, F.R.S., and ironmonger in Thames Street.

of the land on all sides around London. . . . The
vast London, and the frightful number of people
which there crawl in the streets, pay the market-
gardeners many-fold for their outlay."

He is struck by the amount of broken bottle-
glass fixed on walls to hinder climbers, and the
quantity of drink to which it corresponds.

The next day, in the footsteps of Linnæus,
Kalm walked beyond Chelsea to Fulham—"a
pretty town with several smooth streets; all the
houses of brick, very beautifully built,"—crossed
over a wooden bridge,[1] paying a halfpenny toll,
and found that "on the other side of the Thames,
opposite Fulham, there lay a large and tolerably
flat and bare common, which was abandoned to
pastures. It was for the most part overgrown
with furze, which was now in its best flower, so
that the whole common shone quite yellow with
it. In one place only was it cut down for fuel. . . .
In some places we saw Ling."

Putney Heath, at that time continuous with
Barnes Common and Wimbledon Common, must
have been a glorious sight in May.

Kalm is astonished at the number of wigs he
sees in England. "All the labouring folk go

---

[1] The wooden Putney Bridge was built in 1729, and took the
place of the ancient ferry.

through their everyday duties with perruques on the head." "The boy is hardly in breeches before he comes out with a perruque, sometimes not smaller than himself."

It is a comfort that, for three generations, man's everyday dress has become stable, and reasonable. Women's dress in the next century will, no doubt, follow suit.

The day following, Kalm is again at the Physic Garden, and has a long talk with Philip Miller on the vitality of long-buried seeds, and on the heating of greenhouses. He notes that "in the largest orangery in Chelsea Garden the smoke makes six bends in one of the long walls before it escapes." "Orangery" seems to have been a common name for the house into which the "tender greens" were taken for the winter— the "greenhouse."

On May 18th Kalm again spends the morning at the Physic Garden. In the afternoon he pays a touching visit to Sir Hans Sloane. Sir Hans is in bed, aged-looking and rather deaf. He approved highly of Kalm's coming expedition to America, and thought it likely that he would discover many new plants. Kalm writes: "One and all looked upon this man with a particular interest, because he was the oldest of all the learned men now living in Europe, whose names, on account of their writings and learnings, are widely known. We find in the philosophical letters of that learned man, John Ray, several letters which Sir Hans Sloane had written as long ago as the year 1684, together with several of John Ray's answers to them, from which appears what a great insight Sir Hans Sloane had even at that time (aged 29) into all branches of Natural Science."

Kalm pays two visits to Sir Hans Sloane's museum. Among other curiosities he is struck by the way in which Hertfordshire "pudding-stone,"[1] which he must have seen used as boundary stones on Berkhamsted Common, can be polished to make "very handsome" snuff-box lids. Some of these, he was informed, had been sold at an enormous profit in China.

On June 16th, 1748, he is again at the Chelsea Physic Garden, and describes it as "one of the largest collections of all rare foreign plants, so that it is said in that respect to rival the Botanic Gardens of both Paris and Leyden."

In "one room of the Orange-house, in which

---

[1] In "Hertfordshire conglomerate" the matrix in which the pebbles are embedded, like raisins in a pudding, is as hard as the pebbles, and takes a polish.

(*Charles E. Webber, Photo*)

CHELSEA PHYSIC GARDEN LOOKING NORTH
Showing plants arranged according to their places in the natural order

the plants are set in the winter time which cannot bear exposure in the open air, but still do not require any heat, stands Sir Hans Sloane, carved in white alabaster with a scroll of paper in his hand, on a white marble pedestal."

He describes as a "great rarity" Ray's Herbarium[1] of dried plants, with the names of the plants in Ray's own handwriting in a room in the "Orangery," and is enthusiastic over Philip Miller's knowledge of botany and horticulture.

Kalm journeyed from London to Woodford, and on to Little Gaddesden, hoping to get information on English agriculture from William Ellis, who wrote on it, but, according to Kalm, had little practical knowledge.

He is delighted with the beauty of quickset hedges and the shelter they afford, and prefers them to the "dead-hedges" in Sweden. He says: "The beautiful appearance of the country must be ascribed to industry and labour. It resembles

[1] This was presented to the British Museum in 1862. It had been left to the Apothecaries' Society, together with Rand's herbarium, and his own in 1734, by Dr Dale, an apothecary living at Braintree—botanist and naturalist—friend of Ray, Sloane and Petiver. He republished Taylor's *Antiquities of Harwich*, and published a *Materia Medica*. The Daleas (not to be confused with mispronounced Dahlias) were named after him.

one continuous pleasure garden from the many
living hedges there are everywhere." He notes
that the furze and bracken on Berkhamsted
Common make good fuel. He comments severely
on the cold cottages, with no "moss" packed in
the roof, and with grates which allow the heat
of the fire to go up the chimney, and which burn
a large amount of fuel; also on the unoccupied
women, sitting round the fire "without doing the
least thing more than prate."

At last Kalm left for America; and among
the plants he brought back from that country
must have been some of the beautiful heath-like
Kalmias. His plants were nearly lost, for the
captain of his ship—possibly distracted by Kalm's
incessant questioning—ran his ship ashore at
the mouth of the Thames, and had to land his
passengers.

The exciting news that Kalm, with an American
wife, and a heap of new plants, had arrived in
Sweden, cured Linnæus of an attack of gout.

Linnæus' other pupil, J. C. Fabricius, whose
name has already been mentioned—a Dane,
and Professor of Natural History at Kiel—had
studied at Upsala, and had absorbed wisdom
from his teacher.

Fabricius, in 1767, sailed from Holland to

Edinburgh—stayed with his brother at the University—learnt a little English—bought horses, and with his brother rode slowly to London, "gathering plants and insects on the journey." In London he sold his horses "with loss," but gained welcome friends in Sir Joseph Banks and his botanist, Solander.

The next year (1768) he saw Banks off on Cook's first great voyage, and wrote, "It made London appear to me empty."

In 1775 he came to London again. Cook had brought Banks safely home with "innumerable specimens of Natural History"; and Fabricius wrote: "I found plenty of objects to engage my time, and everything that could possibly be of service to me." In 1787 he was again in England.

Fabricius, "the founder of scientific entomology," besides the Physic Garden, found another attraction in Chelsea.

An entomologist, William Jones, who had made a collection of butterflies and moths, and had drawn and described all known butterflies, from his own and other collections,[1] was living

---

[1] The water-colour drawings, known to entomologists as "Jones' Icones," have not faded—some of the half-used cakes of colour have Chinese stamps on them, showing their origin.

Some of Jones' butterflies, too, are well preserved owing to a preparation he put on the bodies.

in Manor Street, close by the Garden. He had been a wine merchant in London—had a house and garden in Chelsea, and was living there a quiet life, devoted to favourite studies.

Faulkner, in *History of Chelsea*, 1829,[1] says that Jones had "realised a handsome fortune." He had also inherited from his parents some property in Sussex; his mother having been a Dawtrey, a family which had lived at Moor House, Petworth, for some centuries, since William Dawtrey (de Hault Rey) founded Hardham Priory—a family now only celebrated for its old tombs in Petworth church.

Faulkner gives a long and pleasant account of Jones. He says: "His learning and abilities were of a most superior order; he was eminently skilled in the Hebrew and Greek languages. ... But it is in the character of a naturalist that he must be principally regarded, having painted from nature 1500 species of butterflies in a most masterly manner, and characterised them in the Latin language, in the Linnæan manner. ... These drawings were so much admired by the celebrated Fabricius that he described from them ... in the last journey alone above 200 new species which he named and published.... These

[1] Vol. II, p. 84.

paintings still exist with his heir[1] in six quarto volumes. . . . He collected personally about 800 species of British Lepidoptera . . . one of the best collections of the day. . . . He painted in oil very successfully, which the pictures that adorned his house testified. . . . "

Jones enjoyed his work, had little ambition, and no wish to publish his drawings. He allowed Fabricius and other naturalists to make what use they liked of them—Donovan[2] and others to copy them. His house in Chelsea became a centre for the London entomologists, who recorded the result of their expeditions.[3] Some of these ento-

---

[1] John Drewitt, whose father, Thomas Drewitt, had married, in 1759, Mary Dawtrey, Jones' cousin. The last descendant of his name has handed over these volumes to Oxford University together with Jones' collection of butterflies. Some of the drawings had been copied by the late Professor Westwood in the writer's rooms at Oxford and had been for many years in the Oxford Museum.

[2] Edward Donovan, the author of several books on natural history containing excellent plates.

[3] In Jones' MSS the great "swallow-tail" butterfly—"swallow-tail fly" as it was then called—now only found in the Fens—is described as being common in the south of England, and the last of another disappearing race of beautiful insects appeared in Chelsea at this time. Sir William Jardine (in the *Naturalist's Library*, 1841, Vol. IV, p. 237), quoting Curtis, says that the Pease-Blossom Moth—*C. Delphinii*—"no less esteemed for its rarity, than for its lovely colours"—which had occurred near Windsor, and of which "the late Duchess of Portland possessed a wing found in a spider's web at Bulstrode," was "once taken by the late Mr Jones in his garden at Chelsea." The faded moth is still in Jones' cabinet.

mologists were descendants of French Huguenots,
who found in natural history, as they did in their
love of gardens, refreshing rest from work.

So Fabricius was often in Chelsea, and although
his devotion to botany was becoming platonic, he
must, on each visit to London, have found time
to go round the Physic Garden, and send news
of it to Upsala. And (as Linnæus and Kalm
before him) he no doubt walked over the wooden
bridge to Putney Heath; but with a butterfly-
net as well as a botanist's box.

At the Physic Garden, Isaac Rand, the Demon-
strator of Plants, was succeeded by Joseph Miller
(no relation of Philip Miller), Apothecary and
F.R.S., author of the *Compendious Herbal*, and
of drawings of plants, now in the Apothecaries'
Library. Joseph Miller was succeeded by Dr
Wilmer; and Wilmer by William Hudson,
Apothecary, F.R.S., and author of *Flora
Anglica*, in honour of whom Linnæus named the
Hudsonias.

Although Sir Hans Sloane had given £150
towards the repair of the greenhouse, he had
bequeathed nothing towards the maintenance of
the Garden. The Royal Society was then asked,
through the Earl of Macclesfield, the President,
to help in maintaining such a scientific institution; but without success.

In 1770, Philip Miller, who was becoming irritable under old age and much study, had to leave the Garden, which for 48 years his genius had made as important as any garden in Europe. He was voted a pension of £60, well deserved. He died the next year, aged 80. Three years before his death the last edition of his great *Gardener's Dictionary* appeared. In it he adopted the Linnean classification because "no other approaches so near to a natural method" and "it is founded on the parts of plants that are most constant." Many years later the Linnean and Horticultural Societies put a monument to his memory in Chelsea church.

At Apothecaries' Hall, meantime, there had been some anxiety. A fire had occurred in Water Lane, close to the Hall. The laboratory which adjoined the Hall now seemed a possible source of danger. So panelling was removed, more brickwork added, and it was decided that, although the furnace might be used for the production of hartshorn, all "vitriol" must be banished—a good ruling.[1]

It was decided, too, that the wharf on the

---

[1] The Apothecaries' Society, which for two centuries and a half had taken the lead in providing pure drugs, has lately, owing to present economic conditions, closed that department of its work.

Fleet should be let, and advertisements for tenders put in the *Daily Journal* and *Daily Post Boy*. There must be good reading in those papers, even in their advertisements.

In 1745 they managed to send £200 to the Lord Mayor's fund for the soldiers who had put down the rising in Scotland.

At Chelsea an important step was then taken. In August, 1771, it was decided to cut down the two cedars, which, as the old maps show, were growing in the middle of the Garden.

It must have required extraordinary courage on the part of the Committee; and there must have been fierce opposition. Even reasonable people would protest against the destruction of such rare and interesting trees; but there would be a far louder outcry from all those who thoughtlessly insist on keeping trees in their wrong places, allow them to overshadow and ruin cottage gardens, to keep precious sunlight and air from windows, and, with little appreciation of the real beauty of trees or landscape, allow them to blot out beautiful views—people who "would not cut a branch," even if it were fretting a Norman window. They must all have protested furiously, as well as those who contribute the

inevitable remark that "although you can cut
a tree down, you can never put it back
again."

It could have been no easy matter to face such
protests. But the Garden Committee were wise.
They had learnt much from Philip Miller. He
had already cut the lower boughs of the cedars,
and it was evident that if 2000 different species
of plants were to be grown, during forty years,
in a garden of less than four acres, it would be
better that four evergreen cedars of Lebanon
should not overshadow the borders.

So two of the Chelsea cedars were cut down,
and the two remaining trees, in exactly the right
place—one on each side of the water gate—
stood up against the sunset sky, in their old age
like stone pines in a Turner landscape—a famous
and a pleasant landmark. Both lived until 1878;
the survivor until 1903.

It is to be hoped that the timber-merchant
who paid £23 for the trees did not lose by his
bargain—probably he did; for the wood he
bought from the Garden Committee was not the
cedar wood from which the old drawing pencils
were made, so pleasant to hold and so easy to cut,
though no doubt the timber-merchant thought
it was. The "cedar-pencil" tree was not a cedar,

but a great juniper[1] growing in Bermuda. The timber from cedars grown in England seems to be of little value.

[1] *Juniperus Bermudiana.* But the old cedar-pencil tree of Bermuda—like the great Spanish-mahogany tree of Cuba—has gone. Less attractive understudies have taken their place, and the Virginian Juniper is now the cedar-pencil tree.

Chapter vii

Death of younger Linnæus.—Offer of Linnean Collections to Banks.—Sir J. Smith purchases Linnean Collections.—Their arrival in Chelsea.—Smith an F.R.S.—His European tour.—His letters to William Jones on foreign natural history collections, plants, and Professors.—The proposed Linnean Society; Jones suggests delay, a quarterly science-breakfast first, later on a society; learned societies sometimes promote "acerbity."—Linnean Collections removed from Chelsea to London.—First Meeting of Linnean Society, seven Fellows attend.—The Demonstrator, past Demonstrators and the Gardener of the Physic Garden among its first Fellows.

FIVE years after his father's death the younger Linnæus, who had little of his father's genius, died unmarried.

The reason he gave for remaining a bachelor throws a lurid light on an unvaccinated Europe.

There were two ladies, either of whom he might have married—one, young and beautiful, had not had small-pox; and so might, at any time, lose good looks—or life; the other had been through that ordeal, and was "disfigured by pock-marks[1]."

So the younger Linnæus remained a bachelor.

It is difficult now to imagine a world where people went about with scarred faces, and where parents were anxiously hoping to see their child-

---

[1] B. D. Jackson, *Linnæus*, p. 317.

ren safely through an attack of small-pox without
much disfigurement, as if they were valuable
puppies waiting for distemper.

At that time a young medical student—afterwards Sir James Smith—breakfasting with Sir
Joseph Banks—heard that the younger Linnæus
had died, and that Banks had been offered, and
had refused, the Linnean Collections.

James Smith—keen botanist—wrote to his
father, a Norwich merchant, and (backed by
Banks) begged for money to buy them. The
father, who trusted both his son and Banks, after
some hesitation, dispatched £500 to Sweden, with
a promise of an equal amount when the collections
arrived in England.

James Smith at once took rooms near the
Physic Garden—Wheeler, the Demonstrator, and
Curtis and Hudson, once Demonstrators at the
Garden, would help in the examination of the
plants. In 1784 the Linnean Collections arrived
in Chelsea—twenty-six cases in all—and were
safely housed in Paradise Row.

Little could Linnæus, when in Chelsea fifty
years before, have dreamed that his treasures
would find their way to that interesting foreign
village—the home of Sloane and Miller.

Banks, whose great collections and library

were, at all times, at the service of fellow naturalists, helped Smith to examine and compare the
many objects of natural history and botany.

Meanwhile Sweden realized the mistake she
had made in allowing collections, which contained
the very specimens to which Linnæus had given
names, his books and manuscripts to leave the
country.

The King of Sweden was away, and returned
too late to rescue them—the ship carrying the
twenty-six packing cases had already sailed.

But England has taken care of them, and
they can be seen by men of science; and when
a scientific nation tried to destroy London, and
bombs fell on each side of the Linnean Society's
rooms in Piccadilly, they were hidden away,
encased in asbestos and steel.

James Smith at once became famous—was
elected F.R.S.—and leaving Banks to continue
the examination of Linnean plants, manuscripts,
and books, in Chelsea, started with a pocketful of
introductions from him on a long European tour.

In a letter to his father from Paris—August 6,
1786—he records a glimpse he had of the un-
happy Court—its last stage, though he was
unaware of it, before the Revolution.

He is not impressed by Versailles—he describes the gardens as "superb but tiresome." He is pleased with the King—Louis XVI—but rather hard on the ladies of the Court.

The daubing of the ladies' cheeks is beyond conception; nature is quite out of the question; old hags, ugly beyond what you can conceive (for we have very inadequate ideas of what an ugly woman is in England), are dressed like girls, in the most tawdry colours, and have on each cheek a broad dab of the highest pink crayon....The King is a pretty good person, rather fat, his countenance agreeable.

In the afternoon Smith is taken to St Germain to see the King shoot.

The game had been all driven together into fields and thickets, around which the people were kept at a distance by soldiers. The King came about three o'clock, stripped off his coat, ribbands, etc., and appeared in a linen jacket and breeches. He was attended by eight pages in the same kind of dress, each of whom carried a gun and one of these guns was always ready charged for the King....Next to these were ten or twelve Swiss Guards...about were principal officers with a physician surgeon, etc....His Majesty went several times up and down the fields killing almost everything he aimed at.

Little could the ill-fated King during that theatrical battue have imagined that Paris was preparing a more deadly one, in which the hunters would become the hunted, and the King, Queen, corrupt courtiers, ill-advisers and all, be surrounded like game to be killed.

Sir James Smith had plans in his mind for a great Natural History Society, with the Linnean Collections as a nucleus. Thomas Marsham, an entomologist, and Dryander, a botanist,—a pupil of Linnæus', and now Banks' librarian—urged him on; and the scheme is more than once mentioned in some letters,[1] written during the tour, to his friend William Jones.[2]

These letters, which show the origin of the Linnean Society, have never been published, so extracts will be of interest.

The first is dated Paris, September 8, 1786, Smith says:

Although so far removed from my friends in England I am anxious that they should not let me slip out of their remembrance and particularly wish...to enquire after the success of your labours in the support of our favourite science....I am not idle in my particular line, which you know is botany, nor have I neglected entomology....I have access to every cabinet and herbarium here, so I have enough to do. I have seen the celebrated collection of drawings of plants and insects begun under Louis XIV and still continued. Some of the plants by

[1] In the possession of the writer.
[2] See pages 89–91.

Robert are very fine, but not better than what we have in England, and I assure you without a compliment the insects are far inferior to yours,[1] in beauty as well as accuracy....I have only seen one thing since I came out that I could much wish you to see, that is the Prince of Orange's Cabinet at the Hague. The insects there are prodigiously fine, the specimens most choice ones, and well preserved, the birds the most rare I ever saw. The *Cabinet du Roi*, at Paris, is paltry.

Although the climate here does not seem better than that of London and the frosts are harder, yet some things flourish better here than with us. Double Pomegranates are very common and extremly fine... and here is quite a profusion of Tuberoses, single and double. I had a stalk of the latter lately, which with its flowers weighed nearly a pound....

I have had a letter from Mr Marsham....He is very anxious about the success of our new Society, which, however, I think had better be kept quite in embryo at present; I have everyday more reason to think it likely to become very respectable, and have it constantly in view. I rely on you to be one of its chief supports. In the meantime I wish you would minute down anything that comes into your mind respecting it, that we may all lay our heads together, and not do anything unwisely or hastily....

Your very faithful fr^d and serv^t,

J. E. SMITH.

[1] These drawings were given to Oxford University in 1925 together with Jones' collection of butterflies.

Lady Smith in *Memoirs of Sir J. E. Smith*, 1832, published the answer—Jones is there described as "well-known in the scientific world, though like other men of superior genius modest and retiring." He advises delay in starting the Linnean Society—he has been "vexed and dissatisfied" with societies. He suggests a "breakfast to select friends once a quarter"—if a society grew out of that, other societies would be less envious, and there would be less of "that acrimony, which exists too much among ingenious[1] people." St Paul's advice, he said, was a good one—"Lay hands suddenly on no man."

Jones evidently disliked the "acrimony" to which scientific associations are liable—an entomological society of which he was a moving spirit had come to a sudden end when some members insisted on the meetings being held on Sundays. A Natural History Society, to which John Hunter the surgeon, and Curtis, belonged, had proved a failure, and Jones must have remembered that two Fellows of the Royal Society had lately fought in the street.[2]

As is well known, societies for the promotion

[1] The old meaning may be rendered as *interested in intellectual or scientific pursuits.*

[2] Page 59.

of science are liable to promote irritability. They contain many active-minded varieties of humanity all differing slightly in experience, manner, temper and temperament, and although they do not end as Bret Harte's geological meeting did, occasional "acrimony" is inevitable.

But Smith was younger, and more ambitious, than his correspondent, and was already collecting foreign members for his Linnean Society.

His next letter is dated Rome, February 19, 1787, he writes:

Here am I in the midst of the Carnival, and so many and such variety of things have I to say to you that I scarcely know where to begin, which is the true reason that I have not sooner written in reply to your obliging favour which I received in Paris....I shall endeavour to profit by your greater experience and judgement... although I myself have had some experience in societies, and some trouble; yet I think on the whole more pleasure and advantage. I cannot help flattering myself that such an association as we have thought of would be useful and respectable in the world, and consequently agreeable to the members of it. Conversations which I have had, in the course of my journey with no small share of the scientific people of Europe, have confirmed me in my opinion that a society for the cultivation of Nat[l]: Hist[y]: *strictly*,[1] is much wanted, and would be ably supported. Where then can it be fixed with

[1] The Royal Society was much occupied with mathematics.

such advantages as in London, amidst the first collections and libraries, and indeed amid the best naturalists that I have known or heard of? I have already engaged many desirable persons to join us whenever we think proper to bring the scheme forward. In the meantime allow me to advise with you at least, and if you approve of the thing when it has acquired something like shape, I hope you will not refuse your assistance in *augendo amabilem scientiam*.

I have great reason to be pleased with my journey thus far....[At Lyons] I saw the cabinet of M^r De Villers containing about 5000 insects of that country....Provence is the most delightful country in every respect that I have yet seen; it agrees with the Republic of Genoa in climate and productions....The *Agave Americana*[1] is naturalized..., and has a fine effect in the landscape. Myrtle, and a vast variety of *Cisti*, with *Pistachia lentiscus, Arbutus unedo*, and many other fine plants cover the hills....

Nothing is more common about Genoa than great hedges of small-leaved double-flowering myrtle, whereas about Florence and Rome they are content with box, bay, and such vulgar things.

I could spend twelve months at Rome with unabated pleasure. Englishmen are perfectly at their ease here, and at Florence, and are extremly beloved, much more so than in France, where indeed they have no reason to complain neither.

---

[1] American Aloe—from which Mexicans make an intoxicating beverage.

Genoa, July 7, 1787.[1]

...I am now quite well and in a week's time shall go to Turin, and from thence thro: Switzerland to Paris, where I shall rejoice to find a letter from you....Am glad I have satisfied you for the present about the Linn[n]: Society; so we may say no more about that matter till we meet, when you shall give your assistance to that project of mine in any manner you please; at least I rely on your counsel....I admire your comparison of the tour of Italy to a journey through life; 'tis just what I have felt....Venice disappointed us: its singularity will always be striking, but nothing there is in a good taste, riches are squandered injudiciously, and dirt deforms everything....I have seen no collections of Nat[l]: Hist[y]: worth mentioning compared with what we have in England....At Bologna are some things and at Pavia more....

We heard Spallanzani[2] lecture; the composition was admirable, but his manner supercilious and affected.... I despair of learning anything about the *Courier Bird*[3] for Mr Latham....

[1] Part of this letter was published by Lady Smith in 1832, but not these extracts.

[2] Celebrated in many departments of natural science.

[3] Dr John Latham, the ornithologist, two years before this, had received the first-recorded British Specimen of the *Cream-coloured Courser*—a rare African bird, which he figured in his book on birds. The skin was afterwards sold for 83 guineas. The bird is so exactly the colour of the desert that on one or two occasions, when the present writer met with it in Egypt, it was invisible until it stood up and ran.

Smith returned to England. The Linnean Collections were removed from Chelsea to London—to Great Marlborough Street—an important street then, for there was no Regent Street—nor Regent.

Close by, in a Coffee-house, the New Society held its first meeting—in 1788—seven Fellows in all—Smith became President—Bishop Goodenough Treasurer—Marsham Secretary, and Dryander Librarian.

In two years' time it had 36 Fellows, 16 Associates, and no less than 50 Foreign Members. Curtis, lately Demonstrator of the Chelsea Physic Garden, and John Fairbain,[1] its Gardener, Martyn (a Chelsea physician, and a Cambridge Professor of Botany), whose translation of Virgil's *Georgics*, with illuminating notes, schoolboys know, Sibthorp, Professor of Botany at Oxford, James Dickson, a florist of Covent Garden, and Beckwith, an entomologist of Spitalfields, were among the first Fellows. Wheeler, the Demonstrator of the Physic Garden, and Hudson, ex-Demonstrator, also joined.

Two years later William Jones became a Fellow; as is seen by a faded receipt for ten guineas, signed by Marsham, among his

---

[1] John Fairbain was Gardener for 30 years.

papers. Sir Joseph Banks became an Honorary Member.

So Sir James Smith started the Linnean Society, which has contributed much to the knowledge of botany and natural history throughout the world.

In order to avoid "acrimony" in the new Society, a rule was made, and was in force for some years, that no remarks should be made on papers read at the meetings!

SIR JOSEPH BANKS (1743–1820), P.R.S.

From mezzotint by W. Dickinson after portrait by Sir Joshua Reynolds

Chapter viii

Sir Joseph Banks as a boy at Physic Garden, fishing with Lord Sandwich; at Eton and Oxford; sails with Captain Cook in 1768; collects plants in Botany Bay; typical old-world naturalist; brings back lava from Iceland for rockery in Physic Garden.— Stanesby Alchorne contributes stones from Tower of London.— Banks and Solander present seeds. — Forsyth, Gardener. — Curtis, Demonstrator. — *The Botanical Magazine.*—Additional tax on Apothecaries.—Botanical excursions.—Thomas Wheeler, Demonstrator; successful teacher; long life.—John Lindley, Professor and Demonstrator, 1835, teaches "natural" system of botany.—"Artificial" system of Linnæus only a link in chain of attempts at a "natural system."—Lindley's energy.—Robert Fortune, Curator, leaves Garden to introduce tea into India.— Expense of the Garden.—Professorship abolished in 1853.— Labourers discharged to reduce expenditure.—Nathaniel Ward introduces "Wardian cases"; attempts to revive Garden.— "Wardian cases" used throughout world.

IN a large house near the south-east corner of the Physic Garden, young Joseph Banks had lived with his mother, and had learnt there the names of plants. He was fond of fishing in Chelsea Reach, and would sometimes pass whole days at his favourite sport with an older and more cunning fisherman, the fourth Earl of Sandwich.

Faulkner—the invaluable Chelsea historian —who, from his little bookshop in Paradise Row, must have often seen Sir Joseph Banks in later life—relates that "even during the night, as the

fish were supposed to bite with a keener appetite,
they" Lord Sandwich and Banks "were accustomed to enjoy their sport in a punt. Their
fishing rods were placed round in due order, and
while they quaffed champagne and Burgundy,
the little bells placed at the extremity of each"
(rod) "gave instant notice of the ravenous barbel,
which, after swallowing the baited hook, ran
away with amazing swiftness, and extended the
silken line to its utmost extremity."

Whether Banks' father would have chosen a
member of the notorious Medmenham fraternity
—the mal-administrator of the Navy—as a companion for his son may be open to doubt, but no
harm, only good, came of the friendship.

An absorbing love of Nature had already taken
possession of the boy. Our thoughtful forefathers
planted our old schools and universities by the
side of rivers. The rich beauty of the flowers
which grew by the Thames at Eton—on banks
unhurt by the wash of launches, untramped by
modern London—had been a revelation to the boy.
He had even carried off to Eton from his mother's
dressing-room a great herbal ("Gerard's" or his
Herbarius, in which Dr Payne found Sir Thomas
More's name) in order to study botany in play-
hours.

And so when at anchor off the Physic Garden
on summer nights, when the little hawk-bells on
the fishing rods were not tinkling, and the exciting
sport of playing fish by starlight had quieted
down, and Lord Sandwich, with nothing to inter-
rupt him but the ripple of the water against
the punt, told stories of his expeditions all round
the Mediterranean Sea, Banks must have been
seized with his great longing to see strange
countries, forests, flowers and butterflies; and to
bring home specimens, drawings, and rare seeds
for the Physic Garden.

So, in spite of having more pocket-money than
is good for youth, and the certain prospect of a
great fortune on coming of age, Banks must have
remembered the fate which, Horace says, awaits
the rich heir, and so set to work to make the best
of life.

As an undergraduate at Christ Church, Banks
established a botanical lectureship for Oxford.
Two years after coming of age he was in New-
foundland collecting plants; and three years
later—on Friday, 26th of August, 1768—through
Lord Sandwich's influence, he sailed with Captain
Cook on the first, and most successful, voyage.

Banks had furnished Cook's ship, and had
engaged a botanist and draughtsmen at his own

expense. He had his reward when the ship anchored and remained for a week in a bay in the "great Southern Continent," and he found himself in a land where queer animals stood on their hind legs like men, among strange trees, shrubs and flowers. There, Banks relates in his journal, the collection of plants became "so immensely large" that he "carried ashore all the drying-paper, spread it upon a sail in the sun, and kept turning it the whole day."

An attempt, meanwhile, was made to pacify natives by putting beads, ribbons and cloth into their huts, but the natives had no use for ornaments or clothes. The presents were left on the ground, and had no more effect than the offer of a silver spoon or inkstand would have on a frightened dog. But the coast was claimed for England, and the spot where the plants were dried was named Botany Bay.

Cook's ship returned to Deal in 1771. The next year saw Banks in Iceland, climbing to the top of Hecla, and bringing back from its desolate slopes a cargo of lava for the Physic Garden.

The adventures and dangers he went through gave him a fellow-feeling for all explorers; so that, when President of the Royal Society, he insisted that the collections and diaries of French

travellers, which our cruisers captured in the
war, should be returned to France unopened—
generosity warmly acknowledged by Cuvier.

He lived till 1820—leaving to the British
Museum his magnificent library and all his
collections.

Sir Joseph Banks was a type of the old-world
naturalist, who, like Gilbert White, Petiver,
Sloane and others, took all Nature as his province.
Such men must have lived a more joyous life than
many a naturalist of the present day, condemned
to search diligently in some little corner of life,
and strain his eyes to learn what his microscope
can teach him.

The old naturalist lived much in the open air
—full of love and admiration for the beauty,
mystery, and infinite variety of Nature. New
forms of life were for ever being presented to
him—none came amiss—tropical flowers and
rare birds, wonderful shells and gorgeous butterflies,[1] minerals, too, and mysterious fossils, and
"curiosities" sailors were bringing from newly-
discovered lands.

Lord Brougham, in a sketch of Banks' charac-

[1] The writer has water-colour drawings, made in 1784, of rare
butterflies in Sir Joseph Banks' collection.

D   8

ter[1] (endorsed by Sir Joseph Hooker), said:
"He showed no jealousy of any rival—no pre-
judice. . . . His house, his library, his whole
valuable collection were at all times open to men
of science." Sir Joseph Hooker speaks of his
"indefatigable exertions" to raise Kew Gardens
"to the position of the first in the world." It was
through Banks' earnest recommendation that
Australia was colonized.

On Sir Joseph Banks' return from Iceland in
1772, a rockery for Alpine plants was made in
the Physic Garden, and the very strangest
company of rocks that ever came together met
in Chelsea.

The blocks of lava, which Sir Joseph Banks had
dug from the lava beds of Hecla, became the bed-
fellows of forty tons of stones from the old Tower
of London, rescued from the road by the Demonstrator of Plants, Stanesby Alchorne, apothecary
and Assay Master of the Mint—stones which had
seen the centuries of tragedy—the heroism and
the villainy—the selfishness and the self-sacrifice
which had gone to the making of England.[2]

[1] *Journal of Sir Joseph Banks*, edited by Sir Joseph Hooker,
p. 31.
[2] In 1772, during alterations in the Tower, ruins of an old stone
wall nine feet in thickness, with Roman coins, were found.—
Wheatley's *London Past and Present.*

ROCKERY OF STONES FROM TOWER OF LONDON (1772), AND LAVA
FROM ICELAND

Catalpa and old Mulberry. Laboratory and lecture room in background

To these John Chandler contributed "a large
quantity of Flints and Chalk;" and (it is to
be hoped much later on) there were added the
inevitable constituents of a London rockery—
broken bricks.

There they are to-day—lava from Hecla, stones
from the Tower—flints from the chalk—round
the basin in the middle of the Garden and on
the bank of the pond. Someone—it may have
been Sir Joseph Banks—contributed two pieces
of brain-stone coral, over which the rock plants
are not allowed to climb.

Stanesby Alchorne, who had rescued the stones
of the Tower, succeeded William Hudson as Demonstrator of Plants. Alchorne gave his services
without salary, presented many new trees, exchanged exotic plants with the Princess Dowager
at Kew, and with the Duke of Northumberland
at Sion House, and received from Sir Joseph
Banks and from Dr Solander (who accompanied
Banks to Botany Bay) a bag of valuable seeds
for the Garden.[1]

William Forsyth was now the Gardener. He
had learnt gardening from Miller, and did justice
to his teacher. During his time (in 1774), although

[1] Alchorne was also public-spirited in making useful experi
ments on himself with the Hemlock of Socrates.

a dam was hastily made at the gate, the Garden
was flooded by high tides to the depth of fifteen
inches. After thirteen years of useful work at
the Physic Garden, Forsyth became Superintendent of the King's Garden at Kensington.[1]

The flower gardens of Kensington Palace, with
Sir Christopher Wren's Orangery in the middle
of them, extended then to the Bayswater Road,
and over the land now occupied by the huge
houses called Kensington Palace Gardens.

The bushes in Kensington Gardens, covered
with small, bright yellow flowers, which cheer
Londoners during the dreary days of March, are
called, after the Gardener, *Forsythia.* Alchorne
added to his services by bringing in as his
successor William Curtis, an apothecary in
Gracechurch Street, a name well known to
entomologists and botanists.

Among those who are susceptible, a love of
botany can be caught—like measles. Curtis[2]
caught it from John Lagg, an ostler at the
Crown Inn at Alton, who knew his *Parkinson*
and *Gerard* well.

[1] Forsyth published in 1802 a *Treatise on the Culture of Fruit
Trees.*
[2] His father was a tanner at Alton, before the rise of industrial
cities, when leather was made which would last a life-time.

Curtis became a devoted botanist, lectured at Apothecaries' Hall as well as at the Physic Garden, and began his *Flora Londiniensis*— plants growing within ten miles of London— with large folio plates of flowers beautifully drawn and coloured, showing their very life and habit of growth—a work too costly to allow of its being continued beyond the sixth volume. He then started the *Botanical Magazine*, which at once became popular. From 1787 it was continued month after month, not only through Curtis' lifetime, but up to the present day.[1] Its author could not have dreamed that it would still be appearing on the bicentenary of Sir Hans Sloane's gift of the Garden.

By 1793, seeds for the Garden had been contributed liberally by Sir Joseph Banks, and also by Sir James Smith, the founder of the "Linnean Society."[2] Seeds and bulbs also came in from St Lucia, Sierra Leone, Port Jackson, Cape of Good Hope, and Madrid. Loam was obtained

[1] In a bookseller's catalogue just received there is a note: "Curtis' *Botanical Magazine*, from 1787 to 1915, with over 8000 coloured plates, £190." A South African tree is named after him *Curtisia*.

[2] Sir James Smith was living in Paradise Row, near the Physic Garden, when the Linnean collections which he had bought arrived in London. See p. 98.

from the Duke of Northumberland at Sion House, and black mould from Lord Spencer, Lord of the Manor of Wimbledon.

But there was more urgent need of funds than of rare seeds; so in 1815 an additional tax on the Apothecaries was again proposed and agreed to. Three years afterwards the barge had to be given up, and the barge-house let to the proprietor of the Swan brew-house.

But instruction in botany went on without hindrance. Since the time of the gallant Thomas Johnson—the editor of Gerard's *Herball*—botanical excursions to see wild plants growing among their natural surroundings had formed an important part of the Apothecaries' training.

It was the duty of the Demonstrator to be the leader of these parties. Five times a year, during the summer, the apprentices and other students met early in the morning. No one was allowed a great coat or umbrella, but each one carried a tin box slung over his shoulder; and there was an attendant, with a larger box for larger plants, following the party.

Sometimes they would tramp through the fields of Islington to the wilder country of Hampstead Heath; sometimes they would cross the river, and wander through Battersea fields, where Fritillaries then grew, as they still grow in the meadows below Oxford, and where that curious plant, the Water Soldier,[1] hid itself in Battersea ditches. Wandsworth, Putney, and Hammersmith were favourite districts for the "herbarizers," for the banks of the Thames were then gardens of wild flowers, and the Inn at Putney a convenient spot for dinner and talk.

There the tin boxes would be opened, and the plants laid out on the table, to be named, classified, and have their medicinal values explained. But many had already fallen from their high estate; the Potentilla was no longer a powerful little flower; Solomon's Seal, graceful as ever, no longer sealed up wounds and broken bones; the sick were not saved by Salvia, and the little Eyebright no longer brought back sight to the blind. A new dynasty was on its way.

Then, in the summer evening, by foot, by river, by coach—counting the days to the next "herbarizing"—the apprentices would find their way home.

Once a year, in July, the Demonstrator conducted an expedition (attended by older botanists only) to find plants growing in the mountains or by the sea. The journeys extended over at

[1] *Stratiotes aloides.*

least two days. The plants collected were exhibited at a meeting, to which distinguished guests were invited. An address was delivered, and a dinner followed, at the expense of the stewards for the year. There is a letter, among the Sloane MSS, from Petiver, asking Sir Hans Sloane to dine at one of these meetings.

Thomas Wheeler was now conducting (and delighting in) the summer excursions. He continued to attend them long after the forty-two years of his demonstratorship were over. A quaint figure—thin and wiry—with bare head, and massive spectacles in front of keen grey eyes, a threadbare coat and long leather gaiters—a man full of kindliness and humour—loved by the students—an inspiring teacher.

Dr Semple,[1] his pupil, says that he was "distinguished for child-like simplicity"; that he "never jested at sacred things, and never uttered a joke which could raise a blush.... His discourses were delivered as he walked, and he never lost an opportunity of saying a wise and instructive word to his young disciples—some of whom even now confess that principles which guided them in mature years were instilled into their minds by this simple-hearted old botanist."

[1] *Memoirs of the Botanic Garden at Chelsea.* Field and Semple, p. 161.

But his appearance was striking. One day when a party of five were returning from a botanical excursion near Maidstone, and Wheeler was on the box with hair blown over his face, laughing and chatting with the driver, and extracting plants from his hat, an excited toll-gate-keeper congratulated the herbarizers on having found the lunatic for whose capture a reward had been offered.

Wheeler was not only a botanist devoted to his master, Linnæus, but a classical scholar; and students had to be careful of their Latin.[1] But conversational Latin was almost dead. Yet a few years earlier, Linnæus had thought it waste of time to learn any language except Latin and his own native tongue. Everyone in Europe with whom he cared to talk—even Philip Miller, the Gardener—could talk Latin.

If Latin had been retained as a universal language, it might not have been necessary to invent a universal gibberish to mitigate the curse of Babel—which falls so heavily on international congresses.

Semple says that from the age of forty until his death, Wheeler "entirely abstained from

---

[1] Anemone, for instance, he insisted on being pronounced with the accent on the *o* when used as a Latin word.

fermented liquors, not from any ascetic feeling," but because he "found himself better without them." He is said to have had "a happy old age," and to have died at 93—a fatal year in the life of botanists. Sloane as well as Wheeler died at that age. Sir Joseph Hooker and Canon Ellacombe only just escaped, and lived beyond it.

In 1829, the Garden Committee decided to throw open the Garden to all students recommended by teachers of medicine and botany, and to give a gold and silver medal annually, as prizes to the two best students—in addition to those already awarded to their own apprentices.

John Lindley, the well-known voluminous writer on botany, became Professor of Botany at the Physic Garden in 1835. He had learnt gardening from his father, an able Norfolk nurseryman, and he was fortunate in being made assistant librarian to Sir Joseph Banks, which must have given him access to books on botany. Later on, he became Secretary to the Horticultural Society, and Professor of Botany at University College. In 1838 he sent an important report to the Government to recommend their taking over Kew Gardens.

Lindley's great work, *Introduction to the*

Natural System of Botany, was dedicated to the Society of Apothecaries. It was founded on the labours of Antoine de Jussieu, and others.

Botanists who accepted the *"Natural"* system, advocated by Lindley, and those who followed the, so-called, *"Artificial"* system of Linnæus, formed opposite camps. The terms are a little misleading to beginners in botany. The antagonism was more in name than in fact. Each system was but a contribution to a natural system—the aim of generations of botanists—a link in a long chain of endeavours to make the truest, and most natural grouping of plants.

Attempts at a natural order had been made even by Gerard, when he wrote of a plant "and its kindes"; by those who divided plants into trees, shrubs, undershrubs and herbs; by Ray and others, who carried natural classification much further. Linnæus, knowing well the complexity of the problem, seized on that important part of a plant—the flower—and made the arrangement of stamens and carpels the foundation of all grouping of flowering plants.

It was an imperfect classification—all botanical classifications are—but it was a well-forged link in the long chain of attempts to form a reasonable sequence of plant life.

Lindley, like Linnæus, added but another stout link to the chain—a chain which was further lengthened when Sir Joseph Hooker (an examiner for prizes at the Physic Garden), brought his wide experience to bear on the subject, and George Bentham his logical faculty and clear use of words (worthy of his uncle, Jeremy Bentham) in separating essential from non-essential resemblance. And the chain will continue to lengthen as the whole life history of plants becomes better known, and the "testimony of the rocks" reveals their long ancestry.

But English botanists were slow to accept the new teaching, and face the great discomfort of changing their opinions. They knew their *Linnæus* by heart, and liked to look on his work as final.

This devotion to a great leader arrested the progress of botany in England.

Linnæus had no such belief in the finality of his system, and (to use the wise words of Professor Patrick Geddes) "the blame of its obstinate and bigoted retention for well-nigh two generations after Linnæus and the elder de Jussieu had departed, must thus, as in so many other historic cases, be ascribed, not to the purpose of the master, but to the blind and indiscriminating reverence of his disciples in adhering to the letter of his writings, at the expense of their general aim and spirit."

At the Physic Garden, Lindley set to work on improvements with energy—lectured to students at the Garden at half-past eight on summer mornings, appealed to Apothecaries' Hall to allow an arrangement of plants on the new system, and encountered inevitable opposition from the old Linnean Curator.

The Curator, Anderson, an early Fellow of the Linnean Society died in 1846. He had been appointed on Sir Joseph Banks' and Sir James Smith's recommendation in 1814. His rough exterior held a good heart. He had done many kind acts, and it was found at his death that a diamond ring given him by the Emperor of Russia had been pledged to help a poor friend.

On Lindley's recommendation, Robert Fortune was made Curator in Anderson's place. Fortune had just returned from an adventurous collecting tour in China, where he had travelled as a Chinaman. He had sent home many new plants—Yellow Jasmine, *Forsythia*, *Weigelia*, and others—now common in English gardens.

Robert Fortune was not long at the Physic Garden. Dr Forbes Royle,[1] who had tried in

[1] See p. 37.

vain to persuade the East India Company to introduce Cinchona trees into India, had obtained the Company's consent to ask the Committee of the Physic Garden for the loan of Fortune's services, in attempting the importation of tea plants. The Garden Committee at once allowed Fortune to give up his post, enter the service of John Company, and undertake this important task.—Hence "Indian tea."[1]

Lindley then appointed Thomas Moore as Fortune's successor.

Perhaps the Garden Committee were not so public-spirited as they appear in parting readily with Robert Fortune; for the cost of improvements in the Garden, when Lindley and Fortune were working together, amounted to £1220. The East India Company could better afford such sums.

And so there arose once more the old question, asked continually for nearly 200 years, How can the Garden bill be paid? It was finally answered by private members subscribing £500, and £700 being paid from the Society's funds. No help

---

[1] In 1851 Robert Fortune successfully introduced 2000 tea plants and 17,000 sprouting seeds in Wardian cases into N.W. India. He published accounts of Tea and Cotton growing in China, and of Rice paper manufacture in Formosa.

came from learned societies, nor from Chelsea landlords, and it was evident that this frequent appeal for funds could not continue.

Under Lindley the long and honourable life of the old Apothecaries' Garden—laden always above the Plimsoll line—had at last reached its climax. Now, in 1853, its decline and fall seemed in sight. Expenses were at once reduced. Lindley's services were dispensed with. Summer lectures ceased. Permanent labourers were discharged. A greenhouse was sold. Tender plants were exchanged for hardy ones, and no fires were lit in the hothouse where Miller had grown some of the first tropical orchids brought to England.

For nearly ten years the Garden remained under partial eclipse. A projected "West London and Pimlico Railway," which would have destroyed it, must have deepened the shadow. Still the Apothecaries continued their prizes in order to encourage botanical students.

Sir Joseph Hooker was now Examiner. In 1858 he reported that the examination was the most satisfactory one he had ever conducted. Charles Hilton Fagge[1] won the gold medal. In 1861, Sir Joseph Hooker reported that Mr Henry

---

[1] Fagge was Physician to Guy's Hospital, and author of one of the most important books of medicine in its day.

Trimen,[1] the gold medallist, was "distinguished beyond all others." The names of William Jenner and Thomas Henry Huxley also appear among the gold medallists.

Both Lord Cadogan and the Royal Society were now asked to undertake the responsibility of the Garden. In both cases the offer was declined.

Then a brave attempt was made to give the Garden new life. The Master and Wardens wrote a letter to members of the Apothecaries' Society to say that "when they reflected how much benefit the Garden had conferred in times gone by, with what pride it had been cherished by their predecessors, and when they found how numerous a body of medical students were still anxious to profit by it—500 having applied for admission during the past summer—they resolved that a vigorous effort should be made to render it efficient."

They thereupon voted £700 and an extra annual grant.

Necessary repairs were carried out, improvements made, and John Ray's valuable collection

---

[1] Trimen, well-known botanist, published with Sir W. Thiselton Dyer, *Flora of Middlesex*. His *Flora of Ceylon* was completed, after Trimen's death, by Sir Joseph Hooker.

of dried plants—each one sewn on a sheet of old hand-made paper neatly labelled and indexed—was handed over to the British Museum for greater safety—a welcome gift. A few years later it was agreed that women students, accompanied by responsible teachers, should have access to the Garden, and that annual prizes should be offered them.

Nathaniel B. Ward, the inventor of "Wardian" cases, who had been Examiner for prizes, was the moving spirit in this heroic rally. He seems to have been a man of gentle and attractive character—from childhood a devoted lover of Nature—and though practising (as his father had done) among the poor in the East of London, was to be seen on fresh, early, summer mornings—before his work began and the world was afoot—among the wild flowers and birds on Wimbledon Common, or Shooter's Hill.

It seemed unfortunate that a devoted botanist should be living amid streets so unfavourable to the growth of flowers; but the misfortune brought about the great discovery of the "Wardian" case.

Ward had buried the chrysalis of a large moth in earth in a wide-mouthed bottle, covered the mouth of the bottle, and placed it in a window.

While waiting for the chrysalis to hatch he

D   9

noticed that little plants began to sprout, and grow in the bottle, which admitted sunlight, but shut out all draughts, and the dry, dusty, dirty air of the street. So a large glass case was made. Plants throve in it. Friends were told of the discovery. Faraday lectured on it at the Royal Institution; and Ward wrote his book on *The Growth of Plants in Closely-Glazed Cases.*

It was a discovery of the greatest importance. Hitherto but few of the plants packed in boxes survived a long voyage. Ward showed that in Wardian cases they were not only unhurt by salt spray, wind, or snow, but that they required no water.

By means of these Wardian cases Chinese bananas were introduced into Samoa and Fiji. Robert Fortune, ex-Curator of the Physic Garden, transported 20,000 tea plants in Wardian cases from Shanghai to the Himalayas. Countless young Cinchona trees crossed over in them from the New World to the Old, and gave quinine to India. Short-lived seeds could be sown in Wardian cases before leaving their own country, and travel, as safely as a child in its cradle, round the world; and the fixed Wardian case in the Physic Garden (full of filmy ferns) shows that these glass boxes temper the harshness of London air.

Ward's friends felt that his discovery ought to have been recognized by the State, but Ward was quite happy without any such recognition, and after his death they found, in his own copy of his work, a quotation from the old *Spectator*: "The consciousness of approving oneself a benefactor to mankind is the noblest recompense for being so." Ward had lived long enough to learn that to have been able to do a good and useful deed is its own and best reward, and to have done the reverse, its own and worst punishment.

This good old naturalist died rather suddenly in 1868. Sir Joseph Hooker (then Dr Hooker) wrote of Ward: "It would be difficult to say which of the many excellent traits of his estimable character was most worthy of imitation, his love of truth, or his appreciation in others of generous qualities far inferior to his own; his unselfish regard for the happiness of those around him; or the absence of all vanity, littleness, or self love....In the memory of those who knew him, he will live as a type of a genial, upright, and most amiable man, an accomplished practitioner, and an enthusiastic lover of Nature."

Chapter ix

THE year 1874 was eventful—not only in the history of the Garden, but in the history of Chelsea. The Chelsea Embankment was then opened.

As far back as 1843 an embankment had been planned; and an offer made to the Apothecaries of a portion of Kew Gardens in exchange for their Garden in Chelsea. The Apothecaries had answered that Kew was too far away for them, and that Sir Hans Sloane's Will put it out of their power—even if they wished it—to make the exchange.

The Chelsea Embankment, in all its newness, sweeping away picturesque gardens, river-stairs, barges, and wharves, was execrated by painters.

Ruskin accepted it for the sake of Carlyle, who liked it and walked on it; and most of those who remembered the smell of the mud at low tide, before the new drainage of London was finished, must have welcomed the change.

But the Embankment was by no means a gain to the Physic Garden.

The trees had been accustomed all their lives to drink twice a day the good Thames water, which at high tide soaked through the earth to their roots; and the total abstinence from it, enforced by the new Embankment, was a sudden change in their way of living, and proved fatal to many.

But harm done by the Embankment is now counterbalanced by good. In all tidal rivers, as an increasing quantity of water is taken from them for the use of an increasing population on the banks, there is a diminishing flow of fresh water to drive back the flood from the sea. Those who now walk over Chelsea Bridge at high tide are probably, without knowing it, crossing salt water—and salt water is not a wholesome beverage for plants.

In the dry summer of 1921 delicate plants at Kew, when watered with Thames water, died. The water was found to be salt, and the great

lake in Kew Gardens which is filled from the
Thames was becoming, by evaporation, a "Dead
Sea." There rain water tanks, long disused, are
being used again.

At Chelsea a most interesting tree survived
the "going dry" of the Garden—a Maidenhair-
tree (Ginkgo),[1] one of the first brought to England.
It was planted against the north wall of the
Garden to protect it from frost in the winter;
but the great Maidenhair-tree in the open lawn
at Kew[2] shows that the precaution was not
needed.

Maidenhair-trees are easily distinguished from

[1] *Salisburia*—often used as an alternative name to the un-
couth word *Ginkgo*—suggests the historic gardens of Hatfield.
But the tree was not named after a Lord Salisbury, but after
Richard Salisbury, Secretary to the Horticultural Society. Just
as the Great Douglas Fir (the Douglas flagstaff at Kew is more
than 200 ft. high) was not called so after a Duke of Hamilton,
but after the intrepid explorer, David Douglas, who was sent
a hundred years ago to America by the Horticultural Society—
discovered more than a hundred new trees and plants—and met
a horrible death by tumbling into a pitfall occupied by a wild
bull.

[2] Sir David Prain informs the writer that the Kew Maidenhair-
tree was planted about 1760. So the Kew tree must originally
have been near—probably against—the stove-house in the Princess
Dowager's garden. A Maidenhair-tree, of about the same age, on
the south side of Holland House is pruned every year, so some of
the leaves, when they begin to turn golden in autumn, are more
than five inches in width.

other trees by the shape of their leaves, which are fan-shaped, like enormous Maidenhair-fern leaves, and they have straight veins as the fern leaves have. The Chinese[1] say that they are like ducks' feet, and give that name to the tree.

The Maidenhair-tree must appeal to all as one of the most wonderful things living on this wonderful earth. It is a link with a bygone time, or which we can have only the faintest conception —a time when the great vegetable world was preparing this planet for man. Trees resembling it were spread over the earth; and the Maidenhair-tree itself was a forest tree in England and Scotland in those far-distant days when the great reptiles dominated the world. Megalosaurs must have crashed through forests of Maidenhair-tree. Iguanodons must have dragged down the young trees for food. The flying dragons, the Pterodactyls, must have rested on its great branches after a raid!

Professor Seward in his interesting book— *Links with the Past in the Plant World*—speaks of it as a "living fossil." But the Maidenhair-tree was a living fossil when the great mammals took the place of the great reptiles—when the mammoth came, browsed on its leaves, and departed.

[1] *Trees of Great Britain.* Elwes and Henry.

It was still more a living fossil in those dim days when the earth was ready for man, whose big brain and cunning hand were destined to give him dominion over all.

This solitary survivor of an ancient race at length disappeared from among wild trees; but it had been rescued and kept alive for the present race of men by the Chinese, who, by some curious instinct, determined to save it, plant it near their temples, and tend it as a sacred tree.

Sir David Prain, the late Director of Kew, who has helped to make those gardens the pleasant resort they are for Londoners, once told the writer that he believed the Maidenhair-tree to be the tree of the "Willow-pattern" plate; and those who compare an old Maidenhair-tree's slender side-branches—covered all the way with small leaves, and turning up gracefully at the end of their downward curve—with those of the so-called "Willow" hanging over the bridge in the picture on the kitchen plate, will at once see the resemblance.

Mrs Bishop, in *Unbeaten Tracks in Japan*, describes great trees of "beautiful *Salisburia adiantifolia*" (Maidenhair-tree) in the wild forest-covered mountains in Yezo—the island of the primitive "hairy Ainos." But the best Japanese authorities say that even these are not wild.

What can be the weakness which now pre-
vents the Maidenhair-tree from holding its own
without the help of man? what the secret of
its extraordinary vitality, which has enabled
it to survive all its fellows, and come down to us
through countless ages from its own strange world?

There are two Maidenhair-trees at present in
the Physic Garden—not old ones.

Young Maidenhair-trees are not picturesque.
Their branches are erect like the artificial variety
of the Black Poplar, the "Lombardy" Poplar,
whose boughs give no shelter from sun or rain,
nor resting-places for birds—"fastigiate" the
botany books call it—with twigs packed close
together like the instrument of punishment in
old-fashioned schools.

The venerable Chelsea tree came to an end
when parish authorities widened the street on
the north side of the Garden. A strip of a few
feet was then taken from the Garden, and with
it the old Maidenhair-tree. The "going dry" of
the Garden had not killed a tree whose race had
survived world cataclysms; but the pavement
in modern London was another matter, and the
tree died.

Two old Mulberry-trees are still flourishing in
the Physic Garden. The accompanying illustration shows one of them.

The Greek Sphinx saw man as a creature which began life on four legs—then stood on two, and finally on three. This old Mulberry doubtless at first required two; then for long years stood on one; and now rests on five.

There are many of these old Mulberry-trees in Chelsea; monuments to repeated attempts— and repeated failures—to cultivate silk-worms with profit.

Chelsea Park, which once extended from the King's Road to Fulham Road, and from Church Street to Park Walk, was taken in 1721 by a "Park Company" and planted with Mulberry-trees as food for silk-worms. Possibly the Huguenot settlers who had market gardens in Chelsea, and silk-looms in Spitalfields, may have suggested the venture. Many of these trees were still to be seen before the ground was prepared for Elm Park Gardens in 1875.[1]

Years before the planting of Chelsea Park, James I had made a Mulberry garden where the King's Road originally began (Buckingham Palace Gardens),[2] and another at Theobald's,[3] and

[1] See p. 29.

[2] In a letter to *The Field* newspaper, December 24, 1921, it is stated that an old Mulberry-tree in Buckingham Palace Gardens has a label: "Planted in 1609 by James I."

[3] The Hon. Lady Cecil, in *London Parks and Gardens*, says that in 1618 a sum of £50 was paid to the head gardener at Theobald's for making a place for the King's silk-worms.

OLD MULBERRY IN PHYSIC GARDEN
—in Winter sleep—leaning on four crutches

had sent peremptory orders all over England to plant Mulberry-trees for silk-culture.

From the time (about 550 A.D.) when Justinian had silk-worms' eggs smuggled into Constantinople, from China, in a bamboo stick, and made silk-culture a royal monopoly, the bright sheen of silk seems to have had an attraction for royalty.

But all attempts, whether by kings or others, to make silk-culture in England profitable have failed. It may be due to England's dull skies, or to the absence of cheap labour, or, possibly, to an attempt to combine the cultivation of silk with that of a pleasant fruit.

Silk-worms will live on the leaves of any kind of Mulberry. Schoolboys know that they will even eat lettuce leaves and in time spin their shrouds of nice yellow silk. But the leaves on which silk-worms thrive best grow on the Chinese Mulberry, which has poor pale fruit. All the old Mulberry-trees which the writer has seen round London have dark fruit. So it seems possible that the Black Mulberry may have contributed to England's failure to produce silk which could compete with the silk of France, Italy, and the East.

On the other hand, it must not be forgotten that the White Mulberry is a more delicate tree,

and so shorter-lived, just as fair-haired children
are less able than dark-haired, dark-complexioned
children to endure life in a London slum—so at
least it appeared to the writer when physician
to a children's hospital years ago. They were
certainly White Mulberries which were *ordered*
to be planted at Hatfield in James I's time.[1]

Another tree, which must have been in existence before the Embankment, is the fine Oriental
Plane near the fern-house—a rare tree in England
for some unknown reason. A native of Asia Minor,
it was planted and prized in old Rome and Greece
for the shade and shelter it gave.

Its deeply-cut leaf, the outline of which schoolboys are told makes a map of the Peloponnesus,
distinguishes it from the common London Plane
—a gardener's variety—planted all over London,
until rows of Planes have become a little mono-
tonous.[2] On the Continent, and in America, Planes

[1] *History of Gardening in England*, by Hon. Lady Cecil.
Lady Cecil kindly reports that there are no White Mulberries
now at Hatfield, but old Black Mulberries, said to have been
planted in James I's time.

[2] The reason—quite unreasonable—given for planting so many
Planes in London is that they shed their bark, and with it the
smoke which has discoloured it—leaving conspicuous white
patches, as if the trees had been slashed. Owing to this habit
of the Plane, it is said that during the occupation of Hyde Park
by troops, an officer, who was not a botanist, censured his men
for the wanton damage he imagined they had done to the trees.

lie under the suspicion of giving off, from their expanding leaves in the spring, clouds of fine down which irritate sensitive throats.

The seed-balls can be seen hanging from the bare branches all through the winter—like marbles or round buttons at the end of a string. They give to the American Plane its name of Button-wood. These buttons serve as labels for Plane-trees when all the leaves have fallen.

There is a fine Oriental Plane at Kew; and also one at Holland House, and at St Ann's Hill, the last one planted by Macaulay's Lord Holland.

The leaf represented on the cover of this book was sketched from one which grew on the Chelsea Oriental Plane. Its outline is not quite that of the Peloponnesus, but the old Greek maps were not made by an ordnance survey.

An old Ilex—an Evergreen Oak—remains unhurt in the south-east corner—not so large, nor so venerable, as the Ilex which, Pliny said, existed in his time in Rome, with a bronze Etruscan label on it. Not even so large as the fine Ilex at the station entrance to Kew, but a tree the Garden may well be proud of. The Ilex aims its long root straight at the centre of the earth, and becomes independent of superficial changes in moisture.

There is a fine Catalpa which must have existed before the Embankment. Catalpas are hardy trees. They can endure the smoke of London, and their large, pale, heart-shaped leaves and white blossoms might be more often seen in London parks. Their timber, too, is useful. In the Middle Temple there is a cabinet made from the wood of a Catalpa which grew in the garden where the Wars of the Roses began.

In Gray's Inn Gardens a venerable Catalpa has a label: "Said to have been planted by Francis Bacon when Master of the Walks in 1598." But the Catalpa, botanists say, was not seen in England until 1728, when it was brought over by Mark Catesby, an explorer and naturalist. Catesby, on an expedition to which Sir Hans Sloane had contributed, found the Common Catalpa in Carolina, near the river Catawba. Hence possibly its name. So the Bacon legend must take its place among others which have grown up round the great Elizabethan like ivy round a dead oak.[1]

But the Gray's Inn tree is well worth a visit.

[1] The Catalpa, too, is a short-lived tree. In *Trees Hardy in the British Isles*, by W. J. Bean, it is said to decline at about fifty years. And Mr A. Heneage Cocks, whose knowledge of country life was always of service to his fellow-members of the Zoological Gardens Committee, told the writer that all the Catalpas planted on his grandfather's property at Marlow are now dead.

With its trunk flat on the ground, it sprawls over the turf in winter (when its leaves have fallen) like some prehistoric reptile.

Catalpas are sensitive to the touch of other trees. The Catalpa in the Physic Garden is upright, for there are no trees near, but most Catalpas in London lean away from their neighbours. The great Catalpa in Manchester Square Garden, driven on one side by a tall Plane, has long required a prop. Years ago the writer lived in Manchester Square, opposite this tree, and could watch from his windows the damage done by the Plane.

When a gale was blowing a long hanging branch of the Plane would swing backwards and forwards like a pendulum, sweeping off the tender shoots on one side of the Catalpa. This had gone on for years, so that the only branches which survived grew on the other side of the tree, and weighed it down in that direction. The Catalpa crouched, like a beaten dog, under perennial flogging by the Plane.

After a time the great swinging bough was removed (not without some protests from other members of the Garden Committee), and the Catalpa is green again on its damaged side, though its balance has for ever gone.

This perpetual pruning of trees by each other must be an important factor in their lives. Once, after a gale at Folkestone, the writer was surprised to see the ground, under a row of tamarisks, covered with fresh twigs as if the hedge had been clipped.

The wind had not blown off the young flexible twigs; they had, by continual rubbing, rubbed off one another.

The chief reason why trees near the coast grow on one side must be—not that the wind pushes them over—but that the twigs on the side exposed to the wind are perpetually jostling and destroying one another, and cannot become branches. So the tree grows in one direction only, and is bent by the weight of the unbalanced boughs.

Japanese Catalpas were later arrivals. The Catalpa in the Physic Garden was struck by lightning some years ago, but the stem escaped. The lightning must have passed through the mass of dripping leaves hanging near the ground. It only destroyed one large branch.

At Holland House there is an American Catalpa, the bole of which, the largest in London, is 8 ft. 10 ins. in girth at five feet from the ground.[1]

Another of the old trees in the Physic Garden

[1] Webster, *London Trees*, p. 169.

(*Charles E. Webber, Photo.*)

CHELSEA PHYSIC GARDEN

Old Persimmon-tree from Southern United States

is a Persimmon from the Southern States of North America, where its fruit ripens and becomes sweet after a wholesome chastening of frost.[1]

Some years ago many were wondering what reason there could have been for giving a horse which had won the Derby (when King Edward, its owner, was Prince of Wales) the name of an American fruit. Bishop Forrest Browne, in his interesting *Recollections*[2], throws light on it. He says: "The Prince told me that the name of his horse Persimmon had nothing to do with the fruit of that name. ... Persimmon was formed from the names of the dam and the sire"—Perdita and St Simon. The name had been given by Lord Farquhar.

On the north wall is a Loquat—Japanese Medlar—growing well; its great crinkled leaves defying London smoke. The Loquat was introduced by Sir Joseph Banks in 1787. There are Loquat trees at Holland House and at Kew, both protected by north walls. Its fruit—like the

---

[1] Since the publication of the first edition of this book, Mr Gilbert Pearson, President of the American Audubon Societies, which are promoting "Nature reserves," and preventing the extermination of rare and beautiful living things of all sorts in the New World, has told the writer that in Florida, even before a frost, the Persimmon is "palatable"—and, among young people, "persimmon pudding" one of the "delights of life."

[2] G. F. Browne, *Recollections of a Bishop*, 1915, p. 341.

Persimmon fruit—does not ripen in the Physic Garden; but most travellers in Mediterranean countries have eaten Loquat tarts. The Barbary Apes on the Rock of Gibraltar used to raid gardens for ripe Loquats.

The Physic Garden Loquat has no claim to be an old one, but the Wistaria on the east wall must not be forgotten. It is old, and may have been brought from China by Robert Fortune[1]; but insignificant when compared with the great Wistarias at Kew[2]—worthy of the festival the Japanese hold in honour of their blossoms. And on the left-hand side of the entrance there is an old, rare Chinese tree with picturesque "pinnate" leaves and pale flowers. The botanists have given it the name of Kœlreuteria paniculata. It is another reminder of the debt European gardens owe to the prehistoric gardens of China.

A Horse-chestnut, too, grows near the entrance—a clumsy tree—not useful—nor graceful, in spite of its white dress during a few days in May. For long years botanists thought that the tree came from Asia. It is only lately that its home has been found in Northern Greece. But

[1] See p. 25.
[2] On the spot where the Princess Dowager's stove-house once stood.

the old Greeks who had keen eyes for beauty of
form, and welcomed the Plane and the Lime,
had no use for the Horse-chestnut—though at
their door.

In England it often occupies the place of a
worthier tree. Happily its heavy leaves fall early,
and let the welcome autumn sun shine on the
dank ground beneath.

There is a Lime, of course—a pleasant and
useful tree—bast-matting and ropes have been
made of its inner bark—its "liber"—for centuries,
and early books were written on it. Romans chose
its timber for shields, and Grinling Gibbons for
the delicate festoons of flowers he carved for St
Paul's Choir, Windsor Chapel, and great country
houses.

Miller, the Curator of the Physic Garden,
taught botanists that the Small-leaved Lime
was the aboriginal Linden tree of England
("Lynden" of Lyte's *Herbal*, 1579).

Lyndhurst must have been a hilly wood (a
hurst) of Small-leaved Lindens before Norman
kings made the heaths, and beech and oak
woods[1] round it, a New and great Forest, within

---

[1] Possibly there was more beech than oak in the New Forest
in Saxon times. A piece of carved wood sent to the writer, cut
from one of the logs on which Winchester Cathedral stood, is

reach of their palace by the transparent waters
of the Itchen; and long before the Dutch Lime
—as Miller called it—became the common Lime
of England.

The Small-leaved Lime does not grow to the
size of the common Lime, and it is not a common tree; but at Rockingham Castle—though
high up, and exposed to the wind—fine old
Small-leaved Limes overlook what is left of
Rockingham Forest, with its great red-timbered
oaks.

beech. If the arrow which killed William Rufus glanced off a
tree, it probably skidded off the smooth bark of a beech, not the
rough bark of an oak.

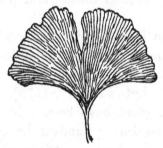

"Duck's foot" leaf of Maidenhair-tree
(*Gingko biloba*)

THE PHYSIC GARDEN LOOKING SOUTH

Old Yew on the right

Chapter x

Old Yew tree.—Yews possibly planted to mark burying-grounds
by Romans.—Cork tree.—Tamarisks.—Ash.—Honey-locust-tree.
—Heliotrope and Mignonette introduced to London at Physic
Garden.

A YEW—more than fifty years old—an uncom-
mon tree in London—grows in the middle
of the Garden. In Webster's *London Trees* this
is said to be one of the largest Yews in the
Metropolis.

It is a well-grown tree. Its low spreading
branches cover the ground, so that not only is
its stem hidden by them, but its roots are protected—as sailors' bare feet are covered by the
expanding bell-shaped legs of their trousers.

This protection of the roots from a parching
sun must be of importance on the shallow soil,
where often the Yew tree grows; while the unwholesome leaves save these low boughs from
destruction by deer or cattle.

Pliny calls the Yew an uncanny, poisonous
tree, and believes the Greek word *toxikon*[1] (poison)

[1] Toxikon itself has an interesting derivation. Literally it means
"pertaining to a toxon" (Greek for bow), and so came to be
used for the poison in which arrows were dipped. When we use
the word "toxic" we are literally speaking of "that which belongs
to a bow."

to be derived from the tree's Latin name *Taxus*. But many more toxic things than a decoction of Yew twigs have been discovered since Pliny's day.

The hardy Yew in the Physic Garden defies London smoke and fog, although its little, close-set leaves see more than one winter, and are not evergreen, but ever black.

It has been suggested that the Yew was planted in churchyards to supply the countryside with long-bows. But the Yew is a slow-growing tree, and if every churchyard contained twenty trees they would not have supplied bows enough. Yew wood, too, has been imported from time immemorial.

Possibly—though the writer is speaking without authority—the custom of planting Yews in churchyards was introduced, as many other customs must have been, by the Romans during the centuries they occupied England.

In Roman cemeteries there was the Cypress. Horace, in his well-known ode to Postumus, regrets the fleeting years, and the coming day when the well-cared-for trees must be left—all except the "unwelcome Cypress."[1]

During the Roman occupation there was no

[1] Horace, Book II, Ode 14.

Cypress tree in England; though the Roman generals may have had boxes made of its iron-like wood. The nearest thing at hand was the Yew.

So when the Roman subaltern ordered his young British slave to plant a tree to mark the new graves outside the garrison town, he would choose a Yew, and probably call it a Cypress; just as a Sallow, which is in blossom about Palm Sunday, is, by a wider stretch of botanical imagination, called a "Palm." If the Irish Yew, with its close, upright twigs, had been known at that time, it would certainly have been chosen, and called Cypress.

So the long-lived, unchanging Yew would take the place of the Cypress, and continue to stretch its strong arms over the ashes of generations of men—even after a wall had been built round the burial-ground, and a Saxon or Norman church had sprung up alongside. And just as the Romans threw sprays of Cypress into the grave at the last farewell, the villagers in England would throw (as they still do) branches of Yew.

A Cork Oak stands near the stove-house—looking like an Ilex. It is difficult to imagine what Apothecaries would have done if there had been no Cork trees. In Spain, Portugal, and across

the Mediterranean there are Cork forests—a
nursery for great soaring birds—eagles, kites,
buzzards—a paradise for naturalists.

Near Bona, in Algeria—a district full of remains of the overwhelmed Roman civilization
—the writer was struck by the ugliness of a
hillside wood of Cork trees with conspicuous
naked stems—by the youth of the trees—and
by the poorness of the bark which was being
carted away—useless except to give a "rustic"
look to a town conservatory. But it is said that
if the bark is removed early in life, the trees get
used to skinning, and in ten years' time give
better cork than those that have not been
stripped.

The thin inner layer of bark along which the
sap runs is left on the trees. But to have their
thick great-coats removed, and to be left with
nothing but very thin vests to protect their trunks
from sun and frost, must delay the growth of the
trees.

Near the Cork tree there is a Cotoneaster,[1]
from the Himalayas, with white flowers in the
summer, and great clusters of red berries in the
autumn, eaten as readily by birds, as cherries by
schoolboys.

[1] *Cotoneaster frigida.*

Near the laboratory door is a well-pruned Bay-tree—*Laurus nobilis*—Victor's Laurel—the true Laurel. Successful poets—poets-laureate—were crowned with its leaves in old Greece.

In England the good old name Laurel has been transferred to the common shrub whose smooth shiny leaves are used by schoolboy-entomologists for killing butterflies—and by rash cooks for giving a prussic acid flavour to puddings.

In this Bay-tree was the nest of an almost entirely white blackbird, which, for years, safely raised a brood there, in spite of London cats, and its own dangerously conspicuous dress.

To the regret of the gardeners, in 1927 the bird was found dead.

It was always an apparition when a white bird dashed out of a bush and across the Garden; for Nature, happily, is sparing in her use of white.

There are few white land-birds; and an accidental one—a white peacock, or a white cock pheasant—is a poor thing compared with those of Nature's own colouring.

A Tamarisk, a little bushy tree with feathery branches, appears in the illustration of the Persimmon. It is a most adaptable plant, varying its habit, like man, with the country in which it is found; growing by rivers, but also drinking

salt water; in the desert far inland where it gives
manna to wandering tribes, but also on the coast
facing south-westerly gales; sometimes a small
hedge, but in a picturesque bay in North Corn-
wall a sturdy tree. There its soft, yielding
branches are being made into lobster pots, while
in Arabia its hard wood is being carved into
camel saddles.

Botany books describe Tamarisk as a recent
immigrant into England. Johnson, when he
wrote his edition of Gerard's *Herball* in 1633,
had not seen it wild. But Sir David Prain told
the writer that he found that a rope, unearthed
with its bucket from the bottom of a Roman
well in Pevensey Castle, was made of the fibre
of Willow and *Tamarisk* bark.

Roman ships coming into Pevensey Bay must
have had coils of spare rope on board—enough
for many wells. But the ships' rigging must have
been of better material—either hemp, or liber—
the inner bark of Lime trees—from which books
were made, and string for Horace's wreaths,[1] as
well as ropes.

So the cord which brought up buckets of
water for Roman cement, when the thick walls
of Pevensey Castle were being built, was, no

[1] "Nexæ philyra coronæ," Horace, Book I, Ode 38.

doubt, home-made—from Tamarisk growing on the Sussex shore.

String round the neck of a Roman jar from Silchester was found by Sir David Prain to be made from Willow-bark—also, no doubt, the result of a home industry.

A volume might be written on String—that necessary part of human outfit—whether used for primitive man's bows and beads, or modern man's millions of brown-paper parcels.

There is a tall Ash—*Fraxinus excelsior*—a variety—*heterophylla, i.e.* "with other leaves." But the "other leaves" are not an improvement on the Ash's own.

The Ash—welcome always—is happily among the many trees which grow well in London.[1] Its wood has always been known to be tough and trustworthy, whether used for Roman spears, or for our aeroplanes—and that of the young plant toughest of all; so that a "ground ash" is used by prefects at Winchester for keeping order in school—as effectually as the kourbash[2] was used

[1] "In Hyde Park and Kensington Gardens upwards of 220 distinct trees are cultivated." Webster, *London Trees*, p. 125.

[2] A piece of untanned hippopotamus hide tapering all its length from the stiff handle to the end of its long flexible lash. Those which used to be sold to tourists in Cairo were ornamented strips of camel skin.

for keeping order during the digging of the Suez Canal.

A Weeping Ash in the S.E. corner of the Garden is over-shadowed by the great Ilex. The Weeping Ash, when it has room, weeps gracefully; its long hanging boughs forming a green tent round the stem, which acts as the tent-pole.

An Ash is made to weep by cutting off its leading shoot and grafting on the wound a sprig of Weeping Ash. There are beautiful Weeping Ash trees in Kensington Square and in many other London gardens. If excessive weeping is checked, and the boughs are supported, they grow to a great length, and make a pleasant canopy—as visitors to the Zoo know.

The Manna Ash, close by, has white flowers in the spring. Its home is Southern Europe, where its sap, as it leaks from slits in the bark, dries into Manna sugar.

There is another Mediterranean tree—a Judas tree—not a large one, but large enough to show its straggling boughs, and rounded leaves, like White-water-lily leaves in miniature. Old tradition says that it is the tree up which Judas,[1] in despair, climbed to end his life.

---

[1] De Candolle considers "Judas Tree" to be "Tree of Judea," but Gerard gives the legend.

Unlike other trees, it has patches of pink blossoms on the bark of branches and stem in early spring.

A Honey-locust-tree in the middle of the Garden has feathery fern-like leaves. Its seed-pods, full of sweet pulp, are kept out of the reach of marauding animals by thorns two or three inches long, and sharp as needles—generally three thorns together—growing from the bark, giving the tree the name of *Gleditsia triacanthus* —a difficult tree for a boy to climb. Its home is the Western States of America.

Among the smaller old trees on the Swan Walk wall is a Styrax, which yields the resin storax, used to relieve coughs since the days of Pliny. And on the same wall is Pomegranate, another old-world shrub, to which the Romans gave the name *Punica*, because they believed that it came to them from Carthage, a Punic (Phœnician) Colony. Its mass of little narrow leaves, shaped like lancet windows, shows that the London winters have not harmed it. Its fruit—(*pomum granatum*—the apple full of grains) reverenced by bygone nations—has now been supplanted by the democratic orange.

The smaller plants are too numerous to mention. But the Heliotrope—sweet Cherry-pie—

must not be forgotten, for Philip Miller introduced
it to the Physic Garden, and to England, in 1759.
Its home is Peru; so it is not the *heliotropium*
of the old botanists, who could not have dreamed
of a New World beyond the Pillars of Hercules
and the Isles of the Blest.

The name means *turning to the sun*, and the *e*
should be pronounced as it is in heliograph and
helium—*he* (the first syllable) should not be
turned into *hel*.

A little fragrant weed too was introduced to
London by the Physic Garden. It had been
brought to Paris from Africa; but was hardly
known in England until Philip Miller, in 1752,
received the seed from Leyden.[1]

An insignificant, colourless flower, with deli-
cious perfume. The French delighted in it, and
called it "The little dear!" London agreed; and
the Mignonette has not changed her name. But
the small head of blossoms has swollen beyond
its original modest size; and sweetness is in
danger of being lost by too great prosperity.

[1] "The seeds were sent me by Dr Adrian Van Royen, the
late Professor of Botany at Leyden....The flowers smell very like
fresh raspberries." Philip Miller in *The Gardener's Dictionary*.

Chapter xi

COMPENSATION for loss of access to the river was paid to the Society, and the money spent on building the present south wall, railings, and iron gates.

The work of the Garden was carried on as before—but with diminishing zest. London had crept far into the country on all sides, and had long since put an end to botanical excursions. Botany had become a less important part of medical training. Expenditure had to be curtailed, and the Garden inevitably relapsed into "winter sleep." Thomas Moore, the Curator, lived there for years among his ferns, and wrote books on them. It inevitably became a neglected Garden, with the damp smell of slow decay. Better far open common, where dead wood can be trodden into earth, and the dead leaves swept by the wholesome wind.

Neither the Royal Society nor the College of Physicians would accept the reversion of the Garden, so, soon after Moore's death, the Apothecaries—finally and sorrowfully—decided to relinquish their trust. In 1893 they handed over the burden they had borne so long to the Charity Commissioners.

It was known that the Garden reverted to Sir Hans Sloane's heirs as soon as it ceased to be used for scientific purposes, and when a notice appeared on the gate to the effect that the Garden was closed, neighbours realized, to their discomfort, that there was a prospect of the Garden becoming a "desirable building site."

Lord Meath, who had done so much to preserve open spaces, summoned a meeting.

Sir William Thiselton Dyer (Director of Kew), Professor Farmer, and others, urged on the Treasury the importance of such a Garden for the use of students.

A departmental inquiry, instituted by the Treasury, reported that the Garden was still well fitted for botanical purposes, and that it was used by students of the Royal College of Science at South Kensington, and of the London Polytechnics and schools. The Trustees of the London Parochial Charities then agreed, on certain conditions, to provide £800 a year—afterwards increased—and the Board of Education £150, for its maintenance. The University of London, Imperial College of Science, Royal College of Physicians, and the Pharmaceutical Society also became yearly subscribers, and a Committee was appointed to superintend its management.

Nine members of the Committee are nominated by the Trustees of the London Parochial Charities, and one each by the Treasury, Lord President of the Council, London County Council, Royal Society, London University, College of Physicians, Apothecaries' Society, and Pharmaceutical Society. Lord Cadogan, as the representative of Sir Hans Sloane, has a seat on it. Sir William J. Collins has for some years acted as Chairman. The first Committee meeting was held in 1899. Mr William Hales from Kew Gardens was appointed Curator.

A strip of the Garden was sold to the Chelsea Vestry for £2,000, to allow a widening of the Royal Hospital Road. This sum, together with £4,050, borrowed by the Trustees, was spent on building the present Curator's house, lecture room, laboratory, and greenhouses. The new buildings were opened in 1902.

There are now three greenhouses, kept at

D

different temperatures; and stove-houses for tropical plants, where weird Pitcher-plants hang their long cups, ready to catch, drown, and devour insects—just as a crocodile drowns and devours creatures which drink at its pool.

Ruskin's sensitiveness, and love of flowers, received a shock on hearing of insectivorous plants. In a letter to the author, speaking of Nature, he wrote:—"I will not be always paying her compliments—the nasty things she turns out!"

But the Butterwort, and the little red Sundew of the moorland, holding midges in its sticky fingers, and the Pitcher-plants of the Tropics, which drown noxious insects, all deserve our thanks.

There are Canary-Island Euphorbias too—with thick, fleshy, upright, leafless, prickly cactus-like branches—unexpected relations of the common little garden weed, Petty Spurge.

At the beginning of our Era, Juba the Second, who had received a good education in Rome, and was reigning in Northern Morocco (Mauretania), wrote a little book about the cactus-like Euphorbia[1], and strongly recommended its acrid, milky juice as a medicine—Euphorbus was his medical

[1] *De Euphorbia herbâ.*

attendant and the plant still has the name given it by King Juba in honour of his doctor.

In Teneriffe these uncanny leafless Euphorbias stand up—spiny sticks, some ten or twelve feet high—among barren fig-trees on the distorted lava slopes—a weird landscape—making an Englishman thankful that he lives in a country where countless æons have gone to the fashioning of the old hills, the rolling chalk downs, the rich red soil, green meadows, and winding rivers, and not on a volcanic island thrown up by our Mother Earth in an angry mood.

A Banana, in another house, is in full bloom (Oct. 1927), and it shows well its magnificent leaves. In a warmer country, in the open air, they would be torn to ribbons by the wind.

In the Garden itself the borders are narrow, and in parallel rows, like printed columns of type; and the plants are arranged according to their places in the latest botanical classification—pages, in fact, of a living book on botany.

Genuine students—there are nearly 3,000 attendances a year—are admitted by ticket. Cut specimens of plants are sent to the College of Science, and to other teaching bodies. More than 1,000 packets of seeds are sent every year to

other botanical gardens. Advanced students work
in the laboratory.

It was there that the riddle, why some peas
acquire wrinkles with age, was answered. At the
present time experiments are being carried on to
discover what fumigation is most fatal to para-
sites of plants—on what stocks Ribston Pippins
are best grafted—what remedy there is for disease
of hops—what alkaloids exist in henbane. Lectures are given in the large room.

Nature study in some form or other is now a
recognized part of education; and much use is
made of the Physic Garden by those engaged in
teaching. A knowledge of botany seems almost
essential to any study of biology. And botany is
the best means of teaching, in a simple way, the
conditions of life and growth in all living things.
Ruskin said truly in one of his early lectures at
Oxford "Real botany is not so much a description
of plants, as their *biography*."

A garden, as Bacon says, is the "greatest refreshment to the spirits of man," and owners of
gardens are fortunate.

But it must not be forgotten that those who
study wild flowers also have their delights; for
the whole world is their garden, and Nature

their gardener. There is a beauty and fitness in living things holding their own among natural neighbours—a charm which cannot be reached by man's art. A bank of bluebells or primroses can be more beautiful than a border of hyacinths or primulas, just as wild creatures in their forests form part of Nature's great picture—not when seen behind bars.

A vulture soaring among mountains is magnificent, but never when captured and tamed.

But it is better to see living things away from their homes than not to see them at all. Plants, too, respond to man's care, and in gardens can outstrip their ancestors in colour and growth. Lessons, too, can be learnt from them which could not be learnt in the wild.

So a botanic garden should be a microcosm of the earth, with its fields, woods, rocks and lakes, where plants can be seen growing, buds opening, and, even in London, bees at work on the blossoms. It widens the views of those whose time has to be chiefly spent in looking through a microscope.

Some knowledge of botany is a necessity to the open-air naturalist; and to the geologist, too, who knows the nature of the soil by the plants growing on it, and learns the history of the crust of the earth, and the temperature of old Continents, by fossil plants in the rocks.

It is an excellent training for the young. It teaches them to look at and love Nature. It leads to drawing, and so increases their power of observation. It teaches them the delicate handling of things, and the accurate use of words.

Those who have been accustomed to observe nature are apt to feel the attraction of botany as years roll by. Gilbert White, after more than ten years of correspondence with Pennant on the *Natural History of Selborne*, wrote to him on its wild flowers.[1] Ruskin at the end of his literary life began a book[2] on botany—with sketches he only could make. Elwes—distinguished traveller and naturalist—spent some of his last years among English trees, and left a monumental record of them.[3]

The love of living plants, with some knowledge of their names and ways, has always been a solace for mankind—a health-resort for deeply occupied minds.

John Stuart Mill, from the time of his discovery of an orange-coloured Balsam on the banks

---

[1] In 1778.

[2] *Proserpina. Studies of Wayside Flowers.* By J. Ruskin, 1879.

[3] *The Trees of Great Britain and Ireland.* By H. J. Elwes and A. Henry, 7 vols. 4to., 1913.

of the Wey—a hundred years ago: in the summer
of 1822—used to turn to a search for wild flowers
with unfailing delight; and must have returned
refreshed by them for more strenuous work.

Arnold of Rugby used to say that the wild
flowers which grow on the Westmoreland Mountains were his "music"—and all must agree with
the great schoolmaster when he wrote[1] that he
"could not bear to see them removed from their
natural places by the wayside, where others might
enjoy them as well as himself."

This love of flowers brings its own reward,
for botanists are among those who know that, in
spite of the rude shocks of life, it is well to have
lived—to have welcomed recurring spring-time
and autumn, and to have seen the everlasting
beauty of the world.

[1] Stanley's *Life of Dr Arnold*, Vol. I, p. 197.

Index

Mill, John Stuart, a botanist, 166
Miller, Philip (1691–1771), appointed Gardener at Physic Garden, 63; publishes *The Gardener's Dictionary*, *ib.*; receives Linnæus at Physic Garden, 68; pensioned, 93; introduces Heliotrope and Mignonette to England, 158
Moore, Sir Norman, on Sloane, p. xvii
Moore, Thomas, (1821–87), Curator of Physic Garden, 126
More, Sir Thomas, his barge, 24; property in Chelsea, 27; his *Herbarius*, 28
Morley, George, Bishop of Winchester, buys Winchester House from Lord Cheyne, 26
Mortimer, Dr, Secretary of Royal Society, 83
Mulberry trees, planted in Chelsea, and Buckingham Palace Garden for silkworms, 137; "Black" and "White" Mulberry trees, 139, 140
Museum, British, founded, 56

Naturalists, old-world, 113
New Forest, its oak and beech woods, 147
Norman, Philip, 27

Oglethorpe, General, founds Georgia, 65

Paradise Row (Chelsea), origin of name, 27, 28, 30
Payne, Dr Frank, 28 n.
Pearson, Gilbert, 145 n.
Pease-Blossom Moth (*C. Delphinii*) in Chelsea, 91 n.
Persimmon, old tree in Physic Garden, 145; origin of the name given to racehorse, *ib.*
Peruvian Bark, *see* Cinchona
Petiver, James (1663–1718), Demonstrator of Physic Garden, 42; his natural history col-

lections, 43; his work on natural history, *ib.*
Petty Spurge, 162
Pevensey Castle, rope of tamarisk fibre found in Roman well, 154
Physic Garden, meaning of name, p. xv; at Oxford and Kew, *ib.* Physic Garden at Chelsea, founded, 20; boundaries of, 20, 21; 25–7 walled in, 31; water-gate made, 32; Dr Herman visits it, *ib.*; cedars planted,*ib.*; described by Evelyn, 36; presented to Apothecaries by Sloane, 46; Garden Committee formed, 62; greenhouses built, 65; expenses, *ib.*; statue of Sir Hans Sloane, 66; visited by Linnæus, 68; by Kalm, 82; two cedars cut down, 94; thrown open to all students of botany, 122; Lindley's improvements, 125; heavy debts, 126; greenhouse, and tender plants sold, 127; Embankment built, 132; Garden deprived of Thames water, *ib.*; trees in, 134–58; passes to Charity Commissioners, 160; Managing Committee appointed, 161; new Curator's house, laboratory, and greenhouses built, *ib.*; arrangement of borders, 163
Physicians, College of, destroyed in Great Fire, 17; their library, *ib.*; Gerard superintends garden, 18 n.; Sloane President of, 53; contributes £100 to Physic Garden, 65
Pitcher-plants, insectivorous, 162
Plague, the Great, 17
Plane-tree, Oriental, 140; London, *ib.*; American, 141
Plants and insects, drawings of, in Paris, 101
Piccadilly, origin of name, 12
Pomegranate, origin of its Latin name, *Punica*, 157
Poplar, Lombardy, 137

Teneriffe, its Euphorbias, 163
Tom Quad, Oxford, copied by Wren at Chelsea Hospital, 21
Tower of London, stones from, for Physic Garden, 114
Trimen, Henry (1843–96), joint author of *Flora of Middlesex*, 128: wins gold medal, *ib.*

Vaccination, Europe before, 97
Vipers' fat, supposed virtue of, 54

Walpole, Horace, trustee of Sir Hans Sloane's Will, 56
Ward, Nathaniel B. (1791–1868), Examiner for prizes at Physic Garden, 129; invents Wardian case, *ib.*; Sir Joseph Hooker's eulogium of, 131
Wardian cases, tea plants introduced into India by means of, 126; and Cinchona trees into India, *ib.*
Warfare, modern, ensures survival of the unfittest, 14
Water Lane, 8; Apothecaries' Hall in, *ib.*
Watts, John, Superintendent of Physic Garden (1682), ex-

changes plants with Leyden, 32; shows Evelyn the Physic Garden, 36
Westbourne, the river, overflows, 21; western boundary of Chelsea, 29
Wheeler, Thomas (1754–1847), joins Sir J. Smith's Linnean Society, 107; 42 years Demonstrator at the Physic Garden, 120; inspiring teacher, *ib.*; Latin scholar, 121
White, Gilbert, 83 n., a botanist in later life, 166
Winchester Cathedral on Beech Wood foundations, 147 n.
Wistaria, 146; at Kew, *ib.*
Woodward, Dr John (1665–1728), expelled from Royal Society, 59; his duel with Dr Mead, *ib.*

Yew, old, in Physic Garden, 149; Pliny's description of the Yew tree, *ib.*; planted in graveyards possibly from Roman occupation of England, 150
Yezo, Island of, Maidenhair-tree forests in, 136

Printed in the United States
By Bookmasters